C/C++ 程序设计

C/C++ Programming

陈宗民 著

清华大学出版社
北 京

内 容 简 介

C++是在 C 语言的基础上发展起来的面向对象的程序设计语言,是 C 语言的超集,所以 C++和 C 语言有密切的关系,C++兼容 C 语言,在很多方面和 C 语言是相通的。

本书以初学者为阅读对象,由浅入深地介绍了 C/C++语言的基本概念,基本语句和语法,基本算法。在此基础之上,重点编写了九个编程实例,以培养读者用所学的编程知识解决实际问题的能力。

全书共分为六部分。第一部分介绍了 C/C++语言的基本概念和基础知识,包括变量和基本的数据类型,基本的程序结构,选择和循环等。第二部分介绍了 C/C++语言面向过程部分的核心特征,主要包括函数、数组和指针。第三部分介绍了面向对象的知识,包括类与对象、类的继承等。第四部分介绍了 C 语言和 C++的文件操作。第五部分介绍了九个编程案例,包括基于机器学习和基于硬件的编程。第六部分是附录,列出了本书一些常用专业词汇的中文解释。

本书采用浅显易懂的英文编写,对概念的描述也有新意,适合中国学生阅读,特别适合双语教学的学生阅读。

本书封面贴有清华大学出版社防伪标签。无标签者不得销售。
版权所有,侵权必究。举报: 010-62782989, beiqinquan@tup.tsinghua.edu.cn。

图书在版编目(CIP)数据

C/C++程序设计 ＝ C/C++ Programming: 英文 / 陈宗民著. —北京: 清华大学出版社,2023.2
ISBN 978-7-302-62648-0

Ⅰ. ①C… Ⅱ. ①陈… Ⅲ. ①C 语言－程序设计－高等学校－教材－英文 Ⅳ. ①TP312.8

中国国家版本馆 CIP 数据核字(2023)第 016028 号

责任编辑: 张　玥
封面设计: 常雪影
责任校对: 李建庄
责任印制: 杨　艳

出版发行: 清华大学出版社
网　　址: http://www.tup.com.cn, http://www.wqbook.com
地　　址: 北京清华大学学研大厦 A 座　　　　邮　编: 100084
社　总　机: 010-83470000　　　　邮　购: 010-62786544
投稿与读者服务: 010-62776969, c-service@tup.tsinghua.edu.cn
质　量　反　馈: 010-62772015, zhiliang@tup.tsinghua.edu.cn
课　件　下　载: http://www.tup.com.cn, 010-83470236
印 装 者: 三河市铭诚印务有限公司
经　　销: 全国新华书店
开　　本: 185mm×230mm　　　印　张: 15.5　　　字　数: 335 千字
版　　次: 2023 年 4 月第 1 版　　　印　次: 2023 年 4 月第 1 次印刷
定　　价: 55.00 元

产品编号: 087043-01

前 言

1. 教材编写初衷

随着时代的发展，我国的对外交流也越来越多，这种交流不仅是经济上的，也有学术和教育方面的。目前，我国很多大学和国外的大学都建立了合作关系，双方联合办学，有"2+2"培养模式、"4+1"培养模式等。这种办学模式往往采用中、英双语教学，双方联合进行培养。这要求学生在国内学习期间能够打下和国外相同的基础，并能够适应可能的国外教学方式。

我校多年来和英国多所大学建立了合作关系，其中 C++一直是一门重要的基础课程。在办学之初，双方通过充分协商，确定了"强基础、重应用"的教学方针。在有限的学时内，重点让学生掌握 C/C++的概念和解决问题的思维方法，放弃了一些不常用的记忆性的细枝末节，着力培养学生用这门程序设计语言解决实际问题的能力。

因为面临语言（英语）和新知识的双重挑战，所以我们一直希望有一本合适的教材。虽然优秀的英语原版教材和中文教材都不少，但无论从内容上还是结构篇幅上都不能完全适合我们的需要。多年来，我们一直使用自编的讲义，中英双方的教师还把自己工作和生活中遇到的一些实际问题编写成一个个案例，来指导学生用程序设计语言来解决这些问题。通过多年的不断完善，我们认为可以将教学经验奉献出来，供具有相同学习需要的教师和学生参考、指正，并共同提高。

2. C/C++，还是 Python

在人工智能及相关技术高速发展的今天，许多大学开设了 Python 语言的课程。这是不是意味着 C/C++已经过时了？是不是应该放弃 C/C++转而学习 Python 呢？我们的体会是这样的，如果把当前的技术热点比喻为一棵枝繁叶茂的大树，Python 可以比喻为郁郁葱葱的枝叶，而 C/C++则可以比喻为树干。Python 短短几行程序可以实现强大的功能，得益于丰富的库的支持，而这些库很多就是用 C/C++编写的。还有，我们知道，C/C++语言虽然是高级语言，同时又被称为中级语言，因为它可以方便地和硬件交互，控制硬件，所以大量的微处理芯片里的程序都是采用 C/C++编写的，比如近年来广泛应用的嵌入式系统，多是采用 C/C++编写的。常用的单片机系统固件（Firmware）的开发，也多采用 C/C++语言。所以，在新的时代，传统的 C/C++语言不仅没有没落，反而更加显现出强大的魅力。

3. 本书的特色

传统上，C/C++作为工科院校的基础课程被广泛学习，也作为后续课程，比如数据结构、Java、Python 等的基础课程。在我校的国际交流学院，这门课程沿用了英方的名称"电子处理系统"。这样命名的用意是：作为电力、电子、电气自动化专业乃至整个工科的大学生，需要了解怎样用程序来分析、处理各种电子信号，处理各种传感器数据，这是一项重要的技能。

当今世界技术的发展趋势之一就是软、硬件相互融合，软件需要运行于硬件平台之上，需要硬件配合来实现。比如人工智能的算法需要运行在一定的硬件平台上，以英伟达 Jetson Nano 为例，只有 70mm×45mm 大小的主板上配有 128 个 Nvidia CUDA 核心的 GPU，可以快速运行 AI 算法，并行运行多个神经网络，同时处理多个高分辨率传感器。

另外，很多硬件，比如单片机芯片等，都需要编写烧录正在芯片内部的程序（固件）才能工作，否则就只是一个空白芯片，这些固件往往使用 C/C++ 语言开发。还有一些硬件，需要先以软件的方式设计，然后再下载、固化到芯片中。比如广泛使用的现场可编程逻辑门阵列（FPGA），需要先使用硬件描述语言（Hardware Description Language，HDL）设计需要的功能，再下载到芯片中。

本书的特色之一，就是精心设计了三个基于硬件的实验，展示程序语言是如何控制硬件的。这三个实验由浅入深，分别是点亮发光二极管（LED）、超声波测距和伺服电机控制。通过程序的控制，一个个发光二极管闪亮起来后，原先看似冷冰冰的代码一下变得鲜活起来，释放出强大的魅力，三十几行代码让一排发光二极管变幻出各种各样的点亮方式，追逐、跳跃、等等，通过修改参数，发光的方式可以变化无穷。学生的学习兴趣和热情一下子调动起来了。

超声波测距实验，即通过不复杂的数学计算和函数的调用可以控制超声波的发射和接收，并计算出超声波遇到物体反射回来所需要的时间，再乘以声波在空气中传播的速度，就可以计算出物体的距离。把手掌放在测声波传感器的前面，前后移动，在计算机屏幕上就可以看到距离的变化。

硬件实验平台采用了风靡世界的开源电子原型平台 Arduino，软件开发环境简单易学，几乎没有学习门槛。在教师的指导下，学生十几分钟就能够使用。接线则采用面包板插接，不需要焊线，非常适合大学一年级学生学习。

三个硬件实验会安排在上半学期，其中前面两个实验安排在前五周。这可以让学生体验到这门语言的魅力，提高学习兴趣，也为后续的学习打下基础。

本书的特色之二，就是由具有多年 C/C++ 教学经验的中英双方的教师联合编写。编写过程中充分考虑到了中国学生的特点，避免采用复杂、冗长、难懂的英文表述，同时放弃了一些平时很少用到的记忆性的知识点。本书融合了双方教师多年的教学经验，抛弃了晦涩难懂的描述方式，特别在概念和思维的讲解上深入浅出、易于理解。针对大学生普遍反映较难的面向对象的概念和方法，结合了生活中常见的例子，一步步地深入，把概念和方法融于实例当中，既降低了学习难度，又让学生在不知不觉中掌握了思路和解决问题的方法。后一阶段的课程设计更加深了学生对概念和方法的理解。这就是"重应用"教学宗旨的体现。

本书的特色之三，就是在排版上对一个页面进行了"二八"比例分割，即一个页面百分之八十的宽度用于正文的排版，另外百分之二十的宽度用于注释正文中出现的生词、关键字等。这样可以方便阅读，避免了学生频繁地查生词，提高了学习效率。

另外，书后给出了三百多个 C/C++专业词汇，便于学生查阅和记忆。

本书作者陈宗民老师具有二十多年的 C/C++教学经验，编写过《计算机信息基础》《C 语言》等教材，曾获上海市优秀教材奖。

编者
2023 年 2 月

目 录

Part 1　C/C++ Fundamentals

Chapter 1　Getting Started with C/C++ ··· 1
　1.1　The C and C++ Programming Language ··· 1
　1.2　Why to Learn C and C++ ·· 2
　1.3　Three Levels of Programming Language ·· 4
　1.4　Interpreted Languages and Compiled Languages ·· 6
　1.5　The Development Tools of C/C++: Compilers and Linkers ······························ 7
　1.6　The Development Cycle ··· 7
　　Chapter Review ··· 9
Chapter 2　Your First C/C++ Program ··· 10
　2.1　Your First C++ Programme on Microsoft VC++ 6.0 ······································ 10
　　2.1.1　Create a Win32 Console Application Project ·· 10
　　2.1.2　How does it work? ·· 18
　2.2　Your First C++ Program on Codeblocks ··· 20
　2.3　Writing a C Program ··· 28
　2.4　Similarities and Difference between C and C++ Program ································ 29
　　Chapter Review ··· 30
　　Programming Exercises ··· 30
Chapter 3　Variables and Operators ··· 31
　3.1　Variables ··· 31
　3.2　Variables and identifiers ·· 32
　3.3　Data Types ··· 32
　3.4　Variables Declaration and Initialization ··· 33
　3.5　Variable Names and Comments ··· 35
　3.6　Operators ··· 38
　　3.6.1　Arithmetic Operators ··· 38
　　3.6.2　Logical Operators ··· 39
　　3.6.3　Bitwise Operators ··· 40

  3.6.4 Relational Operators ·· 41
  3.6.5 Operators Precedence in C/C++ ··· 41
 3.7 Some Much Used Operators in C / C++ ·· 43
  3.7.1 Increment and Decrement Operator ··· 43
  3.7.2 sizeof() Operator ··· 44
  3.7.3 Modulus (%) Operator ··· 45
  3.7.4 Conditional Operator (? :) ·· 46
  3.7.5 comma "," operator ·· 46
 Chapter Review ·· 48
 Programming Exercises ·· 48
Chapter 4 Decision Making ·· 52
 4.1 Conditional Operations ·· 54
 4.2 The if Structure ·· 55
  4.2.1 The if Statement ··· 56
  4.2.2 The if-else Statement ·· 57
  4.2.3 Nested if Statements ··· 58
  4.2.4 if-else-if Statement ··· 59
  4.2.5 switch-case statement ·· 61
 Programming Exercises ·· 64
Chapter 5 Loops ··· 66
 5.1 for Loop ··· 66
 5.2 while Loop ·· 69
 5.3 do...while Loop ·· 71
 5.4 for Loops vs while Loops ·· 72
 5.5 Loop Control Statement "break" and "continue" ···································· 73
  5.5.1 "Break" Statement ·· 73
  5.5.2 "continue" Statement ··· 75
 Chapter Review ·· 76
 Programming Exercises ·· 77

Part 2 Core Language Features

Chapter 6 Arrays ··· 81
 6.1 Declaring an Array 82
 6.2 Initializing an Array 83

- 6.3 Accessing Array Elements ... 83
- 6.4 Array out of Bounds ... 85
- 6.5 Address-of Operator (&) ... 86
- 6.6 Pointers in C and C++ ... 87
- 6.7 Dynamic Array ... 89
- Programming Exercises ... 91

Chapter 7 Functions ... 93
- 7.1 What is a Function ... 93
- 7.2 Library Functions ... 93
- 7.3 User-defined Functions ... 94
  - 7.3.1 Defining a Function ... 94
  - 7.3.2 Parameters and Arguments ... 96
  - 7.3.3 Function Declarations ... 97
- 7.4 Calling a Function ... 98
  - 7.4.1 Passing by Value ... 98
  - 7.4.2 Passing by Address (or Pointers) ... 99
  - 7.4.3 Passing by Reference ... 101
- 7.5 Recursion ... 102
- 7.6 Function Overloading ... 104
  - 7.6.1 Function Overloading by Having Different types of Parameters ... 104
  - 7.6.2 Function Overloading by Having Different Number of Parameters ... 105
- Programming Exercises ... 106

Part 3　Object Oriented Programming

Chapter 8　Strings ... 109
- 8.1 String Assignment ... 111
- 8.2 String Comparison ... 111
- 8.3 String Concatenation ... 112
- 8.4 String Functions ... 113
- 8.5 String Operations ... 114
- Chapter Review ... 116
- Programming Exercises ... 117

Chapter 9　Classes and Objects ... 118
- 9.1 Data Encapsulation ... 118

9.2 Declaring a Class ··· 120
9.3 Defining a Member Function ··· 121
   9.3.1 Getter and Setter ··· 121
   9.3.2 Implementing Member Functions ··· 122
9.4 Creating an Object ··· 124
9.5 Constructors for Initialization ··· 124
9.6 Destructor ··· 126
9.7 Air ticket example for classes and objects ··· 127
9.8 Friend Function of a Class ··· 131
9.9 Operator Overloading ··· 134
Chapter Review ··· 137
Programming Exercises ··· 138

Chapter 10 Inheritance ··· 140
10.1 Implementing Inheritance ··· 140
10.2 Types of Inheritance ··· 142
10.3 Access Modes in C++ Inheritance ··· 144
10.4 Example for Inheritance ··· 145
10.5 Air Ticket Example for Inheritance ··· 146
10.6 Polymorphism in C++ ··· 151
Chapter Review ··· 153
Programming Exercises ··· 155

Part 4 File Operating

Chapter 11 Files and Stream ··· 157
11.1 Types of Files ··· 157
11.2 File Operations in C++ ··· 158
   11.2.1 Creating/Opening a File Using Fstream ··· 159
   11.2.2 Writing Text Files ··· 160
   11.2.3 Reading Text Files ··· 161
   11.2.4 String Streams ··· 163
   11.2.5 Converting to text ··· 164
11.3 File Operations in C ··· 167
   11.3.1 Opening a File ··· 167
   11.3.2 Writing a File ··· 169

11.3.3　Closing a File ···169
11.3.4　Read from a text file ···170
11.3.5　Writing Characters to a File ··171
11.3.6　Appending Data to a File ··172
Chapter Review ···173
Programming Exercises ···173
Chapter 12　Splitting Program into Multiple Files ··································174
12.1　Separate a Program into Multiple Files in C ···174
12.2　Practice in Microsoft Visual C++ 6 ··177
12.3　Separate a Program into Multiple Files in C++ ··179
Programming Exercises ···185

Part 5　Projects for C/C++

Part A　Elementary C/C++ Projects ···187
　Project 1　Logical Gates ···187
　Project 2　Lorry Fleet ···193
　Project 3　Money Class ··198
Part B　Computer Game and Machine Learning Projects ···201
　Project 4　Hangman Game Project ··201
　Project 5　The tic-tac-toe game ··205
　Project 6　Designing a chatbot ···209
Part C　Hardware Based Projects ··213
　Project 7　Blinking a LED and LED chaser ···213
　Project 8　Distance Measurement using Ultrasonic Sensor and Arduino ············218
　Project 9　Servo Motor Projects ··222
Appendix: Vocabulary for C/C++ ···229
References ··233

Part 1   C/C++ Fundamentals

Chapter 1   Getting Started with C/C++

Programming is a useful skill that can open up additional possibilities when undertaking project work later on in your degree, or in your career life. There are many programming languages out there and they all have different strengths and weaknesses. Let's look at where C/C++ came from and why it is useful.

1.1   The C and C++ Programming Language

The C programming language was developed at Bell Laboratories in 1972 by Dennis Ritchie(Figure 1-1), it was derived from an earlier language called "B" (Basic Combined Programming Language, BCPL).

C language was invented for implementing UNIX operating system. In 1978, Dennis Ritchie and Brian Kernighan published the first edition "The C Programming Language" and which is commonly known as K&R C.

In 1983, the American National Standards Institute (ANSI) established a committee to provide a modern, comprehensive definition of C. Then the ANSI standard, or "ANSI C", was completed late 1988. Below are some examples of C being used.

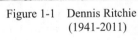

Figure 1-1   Dennis Ritchie (1941-2011)

- Operating system development
- Database systems
- Graphics packages

- Word processors
- Compilers and Assemblers
- Network drivers

字处理软件
编译器和汇编器
网络驱动

The famous UNIX operating system was developed since 1969, and its codes were rewritten in C in 1972. The C language was actually created to move the UNIX kernel codes from assembly to a higher-level language, which would do the same tasks with fewer lines of code.

内核；汇编语言；高级语言

Oracle database development started in 1977, and its codes were rewritten from assembly to C in 1983. It became one of the most popular databases in the world.

甲骨文数据库

Windows 1.0 was released in 1985. Its kernel was mostly written in C, with some parts in assembly.

Linux kernel development started in 1991, and it was also written in C. The next year, it was released under the GNU license and was used as part of the GNU operating system. The GNU operating system itself was started using C and Lisp programming languages, so many of its components were written in C.

发布

C++ came about in 1986 as an Object-Oriented variant of the C programming language. C++ is an enhanced version of the C language. C++ includes everything that's part of C and adds support for Object-Oriented programming (OOP). In addition, C++ also contains many improvements and features that make it a "better C".

面向对象的；变体；增强版本

1.2 Why to Learn C and C++

As a middle-level language, C and C++ combine the features of both high-level and low-level languages.

中级语言
低级语言

It can be used for low-level programming, such as systems programming (operating system, drivers, kernels, networking, etc).

Nowadays, embedded systems are used everywhere, from home appliances to IoT. C and C++ are widely used to write the firmware for an embedded system

嵌入式系统
物联网；固件

as they can directly access to machine level hardware. The most popular development tools for Micro Control Units (MCU) such as Keil and IAR, use C or C++ as its programming language. That is the one of the reasons why after more than four decades from its birth, C and C++ are also most popular programming language.

The syntax of C and C++ are very close to that of human natural language, so the codes of C and C++ are easy to write, understand and maintain. They are general-purpose programming language and can efficiently work on enterprise applications, games, graphics, and applications requiring calculations.

The C and C++ allow a complex program to be broken into simpler programs called functions. So they can be used for large scale software development.

C and C++ are highly portable programming languages which can be easily migrated from Windows to Linux or the other operating systems.

C and C++ languages have formed the basis for many languages including C#, Objective-C, Java, Go, PHP, Python, Perl, Verilog and many more other languages are there.

Figure 1-2 is the top ten programming languages by IEEE in 2020.

Rank	Language	Type	Score
1	Python	🌐 🖥 📱	100.0
2	Java	🌐 ▯ 🖥	95.3
3	C	▯ 🖥 📱	94.6
4	C++	▯ 🖥 📱	87.0
5	JavaScript	🌐	79.5
6	R	🖥	78.6
7	Arduino	📱	73.2
8	Go	🌐 🖥	73.1
9	Swift	▯ 🖥	70.5
10	Matlab	🖥	68.4

Figure 1-2 IEEE top ten programming languages 2020

1.3 Three Levels of Programming Languages

Programming languages can be divided into three levels: low-level languages, middle-level languages and high-level languages. The closer to the machine hardware, the lower level it is.

Machine language, or machine code, is the lowest-level programming language that comprised of binary digits (ones and zeros).

机器语言
二进制数字

Since computers are digital devices, they only recognize binary data. Every character of text, image or video is represented in binary. This binary data, or machine code, is processed as input by the CPU. The resulting output is sent to the operating system or an application, which displays the data visually. For example, the ASCII value for the letter "A" is 01000001 in machine code, but this data is displayed as "A" on the screen. An image may have thousands or even millions of binary values that determine the color of each pixel.

二进制
输入；输出
可视化地
美国标准信息交换码
决定

Assembly language is also low-level language, but is higher than machine code in the hierarchy of computer languages. While easily understood by computers, machine languages are almost impossible for humans to use because they consist entirely of numbers. An Assembly language contains the same instructions as a machine language, but the instructions and variables have names instead of being just numbers. Figure 1-3 is an example of machine language (binary) for the text "Hello World."

汇编语言

指令
变量

```
01001000 01100101 01101100 01101100 01101111 00100000 01010111
01101111 01110010 01101100 01100100
```

Figure 1-3   Machine code for text "Hello World."

Low-level programming languages are difficult to learn, program in and debug, so high-level programming languages are invented. High-level languages are close to natural language, in fact, instructions often look like abbreviated English sentences. High-level programming languages are easier to read, write and maintain, making the process of developing a program simpler and more understandable. Examples of high-level programming languages are C#, Java,

纠错

缩略的

Python, etc. Figure 1-4 is the comparison of high-level languages and the machine code.

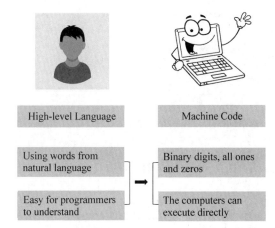

Figure 1-4 The high-level language and the machine code

The middle-level language lies in between the low-level and high-level language. C and C++ are often called middle-level language. By using these languages, the user is capable of doing the system programming for writing operating system as well as application programming.

High-level programming languages are written in a form that are close to human language, enabling the programmers to just focus on the problem being solved. No particular knowledge of the hardware is needed for high level languages programming, the programs written in a high-level language are portable and not tied to a particular computer or microchip.

The main advantage of high-level languages over low-level languages is that they are easier to read, write and maintain. Ultimately, programs written in a high-level language must be translated into machine language by a compiler or interpreter.

Figure 1-5 is the hierarchy of programming languages.

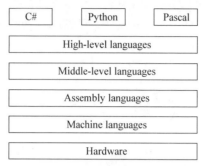

Figure 1-5 The hierarchy of programming languages

1.4 Interpreted Languages and Compiled Languages

解释型语言和编译型语言

Programming languages break into two different families: compiled and interpreted.

The programs of a compiled programming language are translated into machine language by a compiler before being executed.

Compiled languages need a "build" step—they need to be manually compiled first. You need to "rebuild" the program every time when you need to make a change. Examples of pure compiled languages are C, C++, COBOL, Go.

人工编译

Interpreted languages run through a program line by line and execute each command. Interpreted languages are slower than compiled languages. Examples of common interpreted languages are PHP, Ruby, Python and JavaScript(see Figure 1-6).

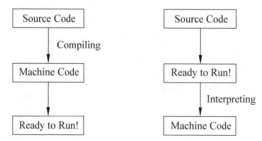

Figure 1-6 The compiled and interpreted Languages

1.5 The Development Tools of C/C++: Compilers and Linkers

> A compiler translates source code into an intermediate form. This step is called compiling, and it produces an object file.

编译器；中间的

Compilers introduce the extra steps of compiling the source code (which is readable by humans) into object code (which is readable by machines). This extra step might seem inconvenient, but compiled programs run very fast because the time-consuming task of translating the source code into machine language has already been done at compiling time. This is not done in scripting languages such as Python and Perl which are interpreted as they run. Because the translation is already done, it is not required when you execute the program. Another advantage of compiled languages such as C or C++ is that you can distribute the executable program to people who don't have the compiler. With an interpreted language, you must have the interpreter installed to run the program on any computer.

源代码
目标代码

耗时的

分发；可执行程序
解释器；安装

> A linker runs after the compiler and combines the object file into an executable program consisting of machine code that can directly be run on the processor.

连接程序

A linker joins the code you write in a source file to both existing libraries of code and the code in other source files you have written and already compiled.

1.6 The Development Cycle

If every program worked the first time you tried it, this would be the complete development cycle:

① *Write* the program…
② …*compile* the source code…
③ …*link* the program, and…
④ …*run* it.

Unfortunately, almost every program, no matter how trivial, may have errors. Some errors cause the *compilation to fail*, some cause the *link to fail*, and some show up only when you run the program (these are often called bugs). Whatever type of error you find, you must fix it, and that involves editing your source code, recompiling and re-linking, and then rerunning the program. This cycle is represented in Figure 1-7 below, which represents the steps in the development cycle.

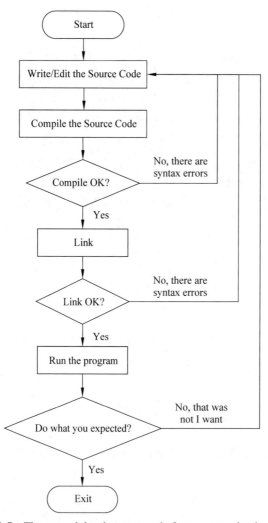

Figure 1-7   The general development cycle for programming in C/C++

If you encounter a problem when compiling, the chances are you have made a syntax error—this means you have typed in code wrongly and will have to go back to your source file and correct it. If your code is compiled successfully, it will build as the linker to find all of the code that your code depends on—if it cannot do this, it will generate linking errors.

语法错误

连接错误

Sometimes a linking error can be fixed by amending your code, while other times it can just be a simple configuration change such as telling the linker which additional directories to look in for code. If the code compiles and links, then it can be run; however, just because code runs, doesn't mean it is right, and for this reason you need to test it by first, understanding what it should do and second relating the output from the program to what the program is doing. If your program doesn't do what you expect, you need to go back and change the source code.

修正

目录；文件夹

Chapter Review

1. What is an interpreted language and a compiled language?

2. What is machine code?

3. What is the difference between a machine-language program and a high-level language program?

4. What is the role of a compiler?

5. What is linking?

6. What is a syntax error?

Chapter 2   Your First C/C++ Program

At this chapter, you will learn how to write your first simple C/C++ Programme. This will give you familiarity with how to create a project and how to compile and link source code that you have written with an external library. Once you have mastered this, we can start to look in more detail at how C/C++ works as a high-level programming language.

程序
项目，工程
外部库

2.1   Your First C++ Programme on Microsoft VC++ 6.0

Traditional programming books begin by writing the words "Hello World" to the screen, or a variation of that statement. This time-honored tradition is carried on here. In this book, Microsoft Visual C++ 6 is used as the complier. It allows you to create many different types of applications. Here we create a Console Applications. A console application is a program that runs inside a DOS window. To create a console application，you need to do the following steps:

控制台应用程序
DOS 窗口

- Create a Win32 Console Application Project.
- Add a source code file to the project.
- Write the program.
- Execute the program.
- Debug the program.

源程序

执行
调试

2.1.1   Create a Win32 Console Application Project

创建 Win32 控制台应用程序

(1) Start Microsoft Visual C++, select "New" from the File menu (see Figure 2-1).

菜单

(2) When the new dialog box opens, choose "Win32 Console Application", then enter the project name in the "Project name" textbox (see Figure 2-2).

对话框

The application provides a default location for saving a project, you can select your own location by pressing the button to the right of the location textbox to

默认的

open the choose directory dialog box.

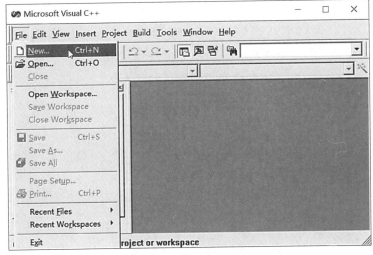

Figure 2-1 Creating a new project

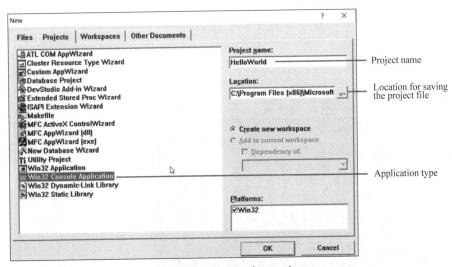

Figure 2-2 New project settings

(3) Press the "OK" button after entering the required information into the dialog box. When the win32 console application wizard appears, select "An empty project", then press the "Finish" button (see Figure 2-3).

When the new project information dialog box appears, select the "OK" button.

We have just created a new win32 console application project. now, we need to add a source file to this project.

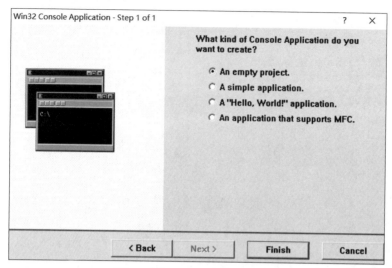

Figure 2-3 New project settings

(1) Selecting "New" from the File menu (see Figure 2-4).

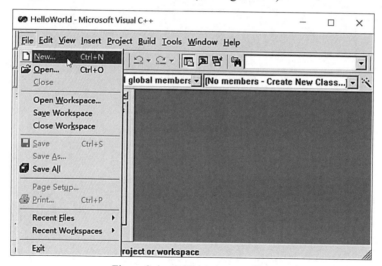

Figure 2-4 Adding a source file

(2) When the new file dialog box appears, select "C++ Source File", type the file name into the "File" textbox, and press the "OK" button (see Figure 2-5).

(3) Write the source code.

Chapter 2 Your First C/C++ Program

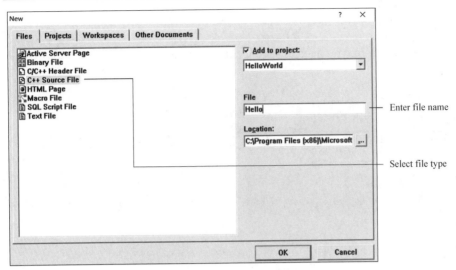

Figure 2-5 Adding source file

Type the source code for your program into the section located in the right side of the window. Figure 2-6 contains the source code for a simple C++ console program that displays the phrase "Hello World" to the screen. Note that the C++ editor automatically selects different colors for C++ reserved words and provides automatic indentation for blocks of code.

编辑器；保留字 缩进

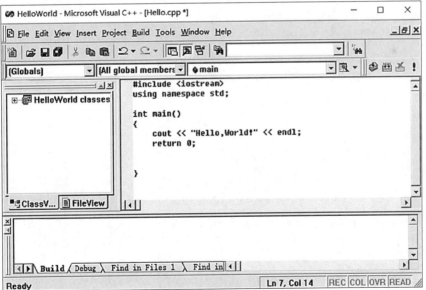

Figure 2-6 "Hello World" source code

13

(4) Save and execute the program

Before you can execute the program, you need to save it (see Figure 2-7).

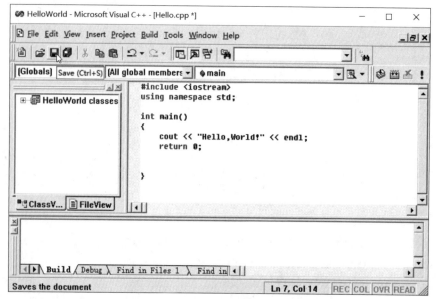

Figure 2-7　Saving the source code

Once the program is saved, you can compile it by selecting the "Compile" option from the build menu or clicking the icon of toolbar (see Figure 2-8).

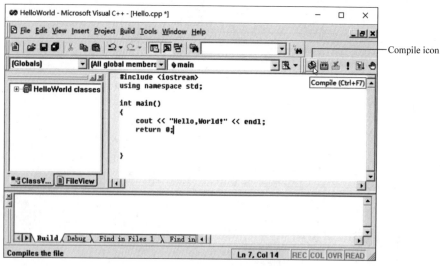

Figure 2-8　Compiling the program

Once the program compiled, the results will be shown in the output window (Figure 2-9). In this example, the program compiled with no errors and no warnings.

输出窗口

警告

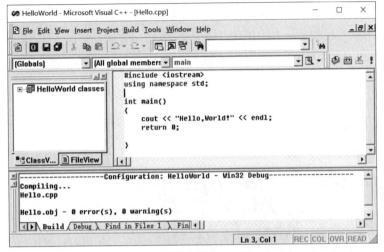

Figure 2-9　Output window of compiling the program

After successfully compiling the program, the next step is to build the executable file necessary to run the program by selecting the "Build" option from the Build menu (See Figure 2-10). In this example, there were no errors or warnings generated by the build process.

成功地编译

生成

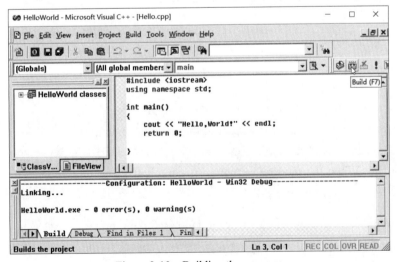

Figure 2-10　Building the program

15

After the build process has been successfully completed, you can now execute the program by selecting "!" icon from the Build menu.

The program results will appear in a new DOS window (see Figure 2-11). Notice that the phrase "Press any key to continue" has been added to the program output by the compiler. This additional code was added to keep the DOS window open until you have had a chance to view the output and press a key on the keyboard. Once a key on the keyboard is pressed, the program stops execution and the DOS window closes.

DOS 窗口

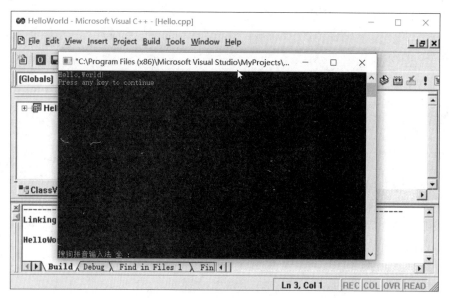

Figure 2-11　The DOS windows

(5) Debug the Program.

调试程序

Often, the program will have errors when you attempt to compile, build or execute it. The program that was successfully executed in the previous example has been modified to include an error—the semicolon at the end of the "return 0" statement has been removed. When the modified program is compiled, an error is displayed in the output window (see Figure 2-12).

修改；分号
语句
显示

You can determine where the compiler found the error by double-clicking on the error message in the output window. This will cause a pointer to appear in the left margin of the source file where the error was encountered (see Figure 2-13).

鼠标双击
信息；箭头
左边的空白处

Chapter 2   Your First C/C++ Program

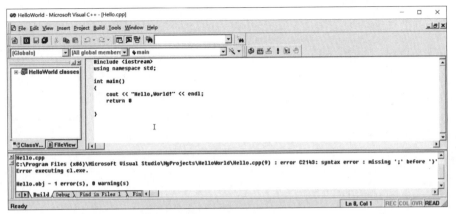

Figure 2-12   Error display window

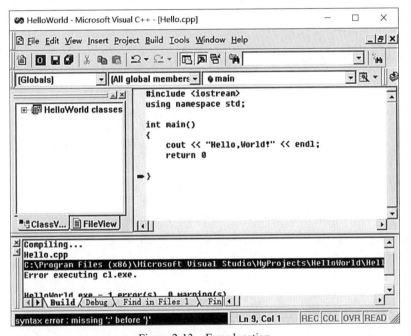

Figure 2-13   Error location

Notice in this case that the pointer is on the line after the line containing the actual error. This occurs when the error induces a compiler recognized fault on a subsequent line. While not always exact, the error pointer, in conjunction with the error description, can help locate errors in the source code. Once the error is identified and corrected, the program needs to be saved, re-compiled, built and executed again.

出现；引起

随后的

找出；重新编译

17

2.1.2 How does it work?

The structure of your project is shown in Figure 2-14. We are using the external library "iostream" in our project. This gives access to the C++ functionality that allows us to output text and numbers to the console window of our computer.

访问
输出文本

```
1  #include <iostream>
2  using namespace std;
3
4  int main()
5  {
6      cout<< "Hello,World!" << endl;
7      return 0;
8  }
```

Figure 2-14   Your first program

Line 1: #include <iostream>

This is a command to the pre-processor which is part of the compiler.

编译预处理器

#include commands allow us to access code that has been built into the C++ language for everybody to use. Here we are accessing the iostream library so that we can use "cout" to display text in the console window.

Line2: using namespace std;

Means we use the namespace named std. "std" is an abbreviation for standard. "cin", "cout" and "endl" are in the "std" namespace. If we don't want to use this line of code, we can use the things in this namespace like this: std::cout, std::endl.

缩写

Lines 4 to 8: This is the main function

Every C++ and C program must contain a main function. The main function is always the first section of code to be executed.

最先执行的程序段

The section of code which belongs to the main function begins line 5 with an opening curly bracket "{" and ends line 8 with a closing curly bracket "}".

You will learn more about functions and how to write your own functions later on.

Line 6: This line of code prints the characters "Hello World!" to the console window

The "cout" object provides the link between your program and the output device of the console window.

The operator symbol "<<" sends the characters "Hello World!" through "cout" to the console window.

"endl" is defined in the iostream library and stands for endline. It causes the "cout" object to start its next output on a new line in the console window.

Line 7: The return statement

This sends information to the computer to tell it that the program has completed successfully.

Some things to note.

Other than the pre-processor command and the curly brackets marking the beginning and end of the main function, every statement ends in a semicolon ";". This character is used in C/C++ to tell the compiler that it has reached the end of a complete code instruction. The C/C++ compiler ignores newline character and blank characters in a source file when compiling.

Try removing the semicolon from the end of line 6 in the code listing, just after the endl and try to compile your source file.

You will find that the program doesn't compile and an error is reported. On my IDE, the error is reported in the following form:

```
helloworld.cpp(7) : error C2143: syntax error : missing ';' before 'return'
```

This is useful because it tells us something about the compilation error and where it has occurred in our program.

The first part of the error message "helloworld.cpp(6) :" tells us which file the compilation error occurred in: helloworld.cpp and the line in our program where the error occurred (6). Some IDEs allow you to click on the error message and the cursor in your editor will jump to the line in the code where your error occurred.

The next part "error C2143: syntax error:" tells us what type of error has occurred in this case. A syntax error means that the compiler cannot translate the symbols in this part of our source file into an instruction in the object file.

语法错误

目标文件

The final part of the error message "missing ';' before 'return'" tells us why it occurred: the compiler expected to see a ";" at the end of line 6.

It is always good to read the compiler errors when your program fails to compile. They will help you work out what you need to change in your source file.

Sometimes when you have a typographical or syntax error in your file, the compiler doesn't generate one error message. It generates 10 or 20 messages.

打字上的；排字的

This can make you sad because it looks as if there were a lot of things wrong with your program and it would take a lot of work to fix the errors. If this happens, the thing to do is to fix the first error on the list, then try compiling your program again. Often you will find that correcting the first error will solve many of the other problems in your compilation.

改正错误

2.2 Your First C++ Program on Codeblocks

CodeBlocks is an open-source, cross-platform (Windows, Linux, Mac OS), and free C/C++ IDE. It supports many compilers, such as GNU GCC and MS Visual C++.

开源；跨平台的

Installation of CodeBlocks

As an open source application, CodeBlocks can be downloaded freely from its official website http://www.codeblocks.org. Identifying the correct package is the first essential task, because there are couple distinct packages available. For Windows, download the one named "codeblocks- XX.XX-mingw-setup.exe" (see Figure 2-15).

安装 CodeBlocks 安装包；不同的安装包

For Linux and Mac OS(see Figure 2-16), download the version corresponding to your OS.

操作系统

Chapter 2   Your First C/C++ Program

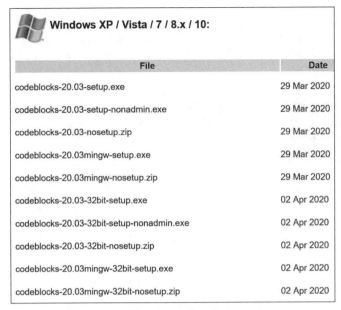

Figure 2-15   Choosing an installation package

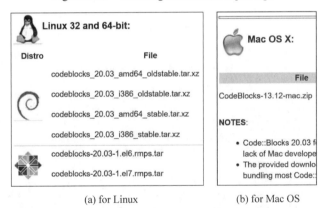

(a) for Linux   (b) for Mac OS

Figure 2-16   Installation packages for Linux and Mac OS

Once you have downloaded the correct package, run the downloaded executable file, and follow its instructions. The default options are fine. The CodeBlocks development environments startup window looks like Figure 2-17.

可执行文件
默认的选项

To start a new project, click "Create a new project" on the screen. Here, you will encounter with a huge list of predefined project templates, as in Figure 2-18. Go ahead and select "Console application". This will allow you to write a program for the console.

工程模板

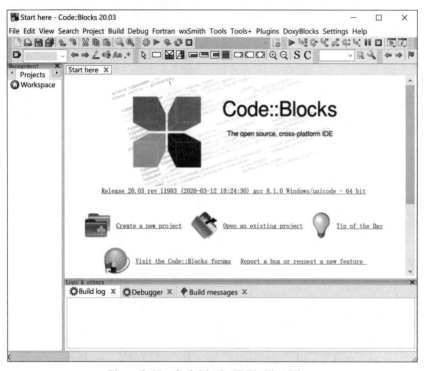

Figure 2-17　CodeBlocks IDE's First View

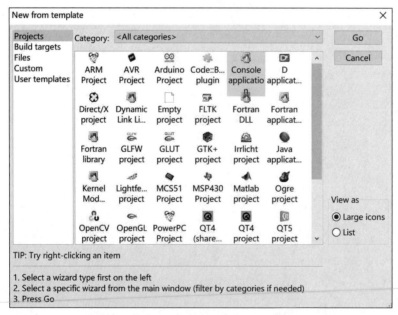

Figure 2-18　CodeBlocks project selection window

Like VC++ 6.0, console application will allow you to write a program on the console. Other applications are for developing more advanced types of applications. After selecting console application, click on the "Go" button to begin using the console application wizard.

Figure 2-19 window allows you to choose the language that you will use. Select the language as C++, then press "Next" Button.

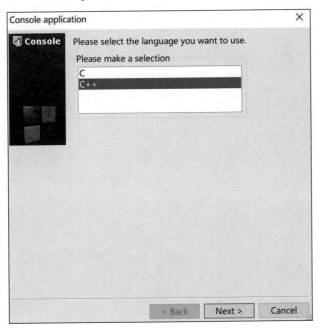

Figure 2-19   Selection of language

Here you need to fill in the project title. You will notice that the project filename automatically becomes the same name with the project name. If you wish, you can change the filename. To specify the location of the folder to contain the project, click on the "..." button and browse to a folder on your drive to store the project. Then press "Next" Button (see Figure 2-20).

文件夹
浏览；硬盘

Figure 2-21 will be the compiler selection window. You can choose a compiler that you use. GUN GCC compiler is the default selection. Then press "Finish" Button.

编译器选择

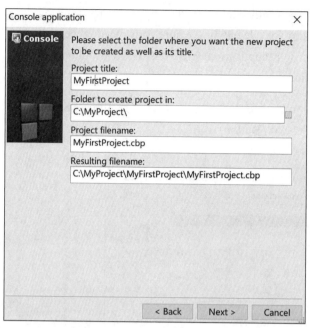

Figure 2-20   Selection of project name and folder

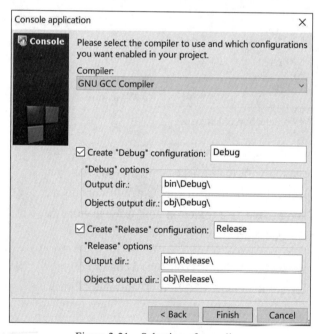

Figure 2-21   Selection of compiler

The system will then return to the "MyFirstProject" window and you are ready to write your program (see Figure 2-22).

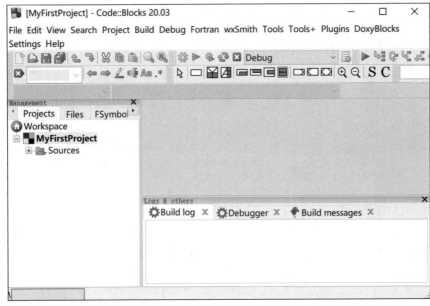

Figure 2-22 Project window

In the Management area of the screen (Left, Shift-F2 toggles the Management display), you will see the files that are part of the project in the projects tab. To see the source files, click on the "+" to expand the Workspace and its subdirectories.

切换
标签页
工作空间

Under "sources", there is a file called main.cpp, which is automatically created for you when you build a console application (see Figure 2-23).

You can try to build that source code by choosing "Build"→ "Build from the menu", or by clocking the yellow gear icon "Build" on the toolbar. This single step does several things (see Figure 2-24).

工具栏

The build log tab in the logs part of the window displays the results of building the project. You see a few lines of text (see Figure 2-25).

创建记录标签页

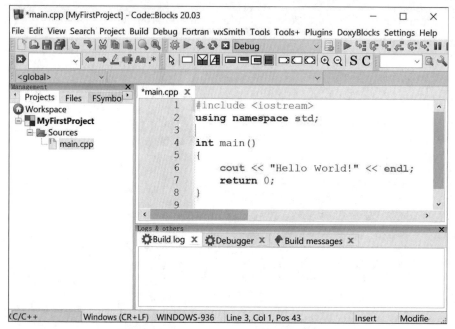

Figure 2-23　Default code of main.cpp file

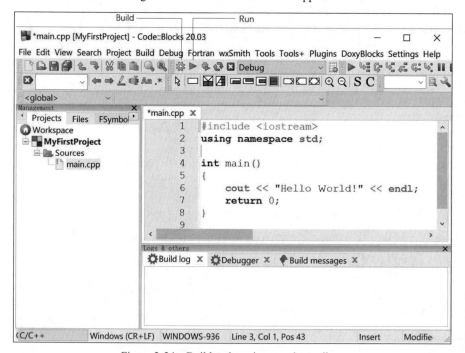

Figure 2-24　Build and run icon on the toolbar

Chapter 2 Your First C/C++ Program

```
main.cpp X
1  #include <iostream>
2  using namespace std;
3
4  int main()
5  {
6      cout << "Hello World!" << endl;
7      return 0;
8  }
```

```
Logs & others
Build log X  Debugger X  Build messages X
-------------- Build: Debug in MyFirstProject (compiler: GNU GCC Compiler)---------------
g++.exe -Wall -fexceptions -g  -c C:\MyProject\MyFirstProject\main.cpp -o obj\Debug\main.o
g++.exe  -o bin\Debug\MyFirstProject.exe obj\Debug\main.o
Output file is bin\Debug\MyFirstProject.exe with size 74.76 KB
Process terminated with status 0 (0 minute(s), 0 second(s))
0 error(s), 0 warning(s) (0 minute(s), 0 second(s))
```

Figure 2-25 Building information

For this demo project, you see the text within the summary that indicates zero errors and zero warnings. For your own projects, when errors appear, which is often, you fix them. The error messages help in that regard.

Building a project is only half the job. The other half is running the project, which means executing the completed program from within the IDE.

To run the current project, choose "Build"→"Run from the menu", or click the green arrow icon "Run" on the toolbar. You see the terminal window appear, listing the program's output, plus some additional text (see Figure 2-26).   终端窗口

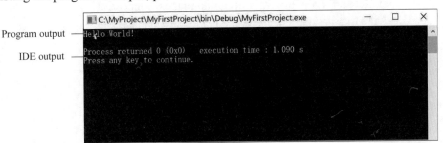

Figure 2-26 The output window

The program output appears in the top part of the terminal window.

27

The last two lines are generated by the IDE when the program is run. The first line of text shows a value returned from the program to the operating system, a zero, and how long the program took to run (1.090 seconds). The "Press any key to continue" prompt means that you can press any key to close the window.

返回

提示

2.3 Writing a C Program

We have just created a C++ Program to output "Hello World!" to the console. Let us rewrite this program in C. In fact, a C program is quite similar in most parts of a program with that of a C++ one (see Figure 2-27).

```c
#include <stdio.h>

int main()
{
    printf("Hello World!\n");
    return 0;
}
```

Figure 2-27 "Hello World" program written in C

Like C++, all valid C programs must contain one main() function. The code execution begins from the start of the main() function.

The printf() function is a library function to send formatted output to the screen. The function prints the string inside quotations.

To use printf() function, stdio.h header file need to be included.

"\n" is a new line character. C and C++ programming languages support a concept called escape sequence. When a character is preceded by a backslash (\), it is called an escape sequence and it has a special meaning to the compiler (see Table 2-1).

转义字符

Table 2-1 Some escape sequences which are available in C and C++

Escape Sequence	Description	Representation
\'	single quote	byte 0x27 (in ASCII encoding)
\"	double quote	byte 0x22 (in ASCII encoding)
\?	question mark	byte 0x3f (in ASCII encoding)
\\	backslash	byte 0x5c (in ASCII encoding)

续表

Escape Sequence	Description	Representation
\a	audible bell	byte 0x07 (in ASCII encoding)
\b	backspace	byte 0x08 (in ASCII encoding)
\f	new page	byte 0x0c (in ASCII encoding)
\n	new line	byte 0x0a (in ASCII encoding)
\r	carriage return	byte 0x0d (in ASCII encoding)
\t	horizontal tab	byte 0x09 (in ASCII encoding)
\v	vertical tab	byte 0x0b (in ASCII encoding)

The following code demonstrates the use of escape characters.

```
1  #include <stdio.h>
2  int main()
3  {
4      printf("This\nis a test.\nShe said, \"How are you?\"\n\a");
5      return 0;
6  }
```

Output is below.

```
This
is a test.
She said: "How are you?"
```

"Hello World" program with C and C++ is below.

```
#include <stdio.h>              #include <stdio.h >

int main()                      int main()
{                               {
    printf("Hello World!\n");       printf("Hello,World!");
    return 0;                       return 0;
}                               }
```

Figure 2-28 "Hello World" program with C and C++

2.4 Similarities and Difference between C and C++ Program

Similarities is below.

(1) Both C and C++ have a similar syntax;

(2) Code structure of both C and C++ are same;

(3) Nearly all of C's operators and keywords are also present in C++ and do the same thing;

(4) The compilation of both the languages is similar;

(5) Basic memory model of both is very close to the hardware.

Table 2-2 is the Comparation of C and C++.

Table 2-2 Comparation of C and C++

C	C++
Header file: #include <stdio.h>	Header file: #include <iostream>
Do not use namespace	Use namespace
Use printf() for output to console	Use cout for output to console
Use scanf() for input from keyboard	Use cin for input from keyboard
C is a subset of C++	C++ is a superset of C
C supports procedural programming	supports both procedural and object oriented programming paradigms

Chapter Review

1. What is an IDE?

2. How to create a win32 console application?

3. What does the following statement do?

   #include namespace std;

Programming Exercises

1. Write a C program that displays your name and your university name on the screen.

2. Rewrite the code of the first question in C++.

Chapter 3 Variables and Operators

Input, process, output…input, process, output…input, process, output…   输入；处理；输出

It's all computers do. Think about it: no matter what software you're using, that's all that's going on. Input, process, output (IPO), is a computer model that all processes in a computer must follow.

For example:

- Input e.g. read from network/disk/database/hardware, accept user input.
- Process e.g. FFT, sum, product, random shuffle.
- Output e.g. write to network/disk/database, flash lights, change display, respond to user input.

Before we can do any processing, we need to have data. And once we get that data we need to hold onto it in some way. How do we do this? We use variables.   变量

3.1 Variables

A variable is a "container" in which a data value can be stored inside the computer's memory.   值
内存

The stored value can be referenced by using its name later in the program.   引用

A variable looks like a bottle in somewhat. A bottle can be used to hold different liquids such as water, a variable can hold or store the value of different data type.

Let's imagine that if you were asked to remember a number: 5. What happened in your brain? You stored this value in your memory. Then, if you were asked to add 6 to the number, you should be retaining the numbers 11 (that is 5+6) in your memory.

The whole process described above is a simile of what a computer can do with two variables. That can be expressed in C/C++ with the following set of statements:

```
a = 5;
b = 6;
c = a + b;
```

Variables are the lifeblood of software—the medium through which data travels all around your programs. The operations described in this chapter demonstrate how to store, process, assign, manipulate and transfer data through the use of variables.

血液；媒介

赋值；操作；传输

3.2 Variables and identifiers

标识（zhi）符

Each variable needs a name that identifies it and distinguishes it from the others. In the following chapters, we will give name to other programming elements, such as functions, classes, etc. All these elements have a common name—identifier.

An identifier is a name that is assigned by the user for a program element such as variable, type, template, class, function or namespace.

赋予，给予

When naming an identifier, we should obey the naming convention. A naming convention is a set of rules for choosing the character sequence to be used for identifiers.

命名约定

(1) Only alphabets, digits and underscores are permitted.
(2) It must begin with either a letter or an underscore.
(3) Key words cannot be used as a name.
(4) Upper case and lower case letters are distinct. C/C++ is case-sensitive.

字母；数字；下画线

关键字
区分大小写的

3.3 Data Types

数据类型

Each variable has a specific type, which determines the size and layout of the variable's memory; the range of values that can be stored within that memory.

特定的类型

There are following basic types of variable in C/C++ in Table 3-1.   基本类型

Table 3-1 Data Types in C/C++

Type	Description	Size
bool	Stores either value true or false	1 Byte
char	Storage of individual text	1 Byte
int	Storage of numbers without a fractional part	4 Bytes
float	A single-precision floating point value	4 Bytes
double	A double-precision floating point value	8 Bytes

布尔型
字符型
整型
浮点型
双精度型

Please notes:

- The character type "char" can be used to store numbers, but is generally used to store characters, e.g. 'a', 'H', '?'. The character is stored as a number which specified by ASCII.   字符型   ASCII 码

- The range of values that can be stored in the int type is limited +2147483647. And although this seems to be a very large number, the population of China is currently (2020) 1366000000. So you would not be able to use an int to store the total savings of the population of China, for example.

- The float type allows a wide range of values to be stored, but there are limits in precision, that is how accurately a number can be represented in a floating point format. A float will be able to accurately represent a number to 9 significant places.   格式   有效数字位数

- The bool type is used when looking at logical expressions where there are only two possible results True and False.   逻辑表达式

The data types described above cover nearly all of the basic data types in C/C++. However, there is always an exception and in C/C++ this is void. The void data type represents nothing and this will make little sense to you until we discuss functions.   空类型   函数

3.4 Variables Declaration and Initialization

变量的声明和初始化

A variable must be declared before it is used.

A declaration specifies a type, and contains a list of one or more variables of that type as follows:

variableType variablelList;

Here, "variableType" must be a valid C/C++ data type including char, int, float, double, bool or any user defined object, etc, and "VariableList" may consist of one or more identifier names separated by commas. Some valid declarations are shown here:

```
int i, j, k;
char c, ch;
float f, salary;
double d;
```

Variables are initialized (assigned a value) with an equal sign followed by a constant expression. The general form of initialization is:

variable_name = value;

Note that the use of the '=' sign is for assignment rather than equivalence. Variables can be initialized (assigned an initial value) in their declaration. The initializer consists of an equal sign followed by a constant expression as follows:

type variable_name = value;

For examples: int number = 12;

Let's look at this line of C/C++ code in more detail:

int: this tells the compiler what the type of the variable is. The compiler needs to know about the type of the variable because the amount of memory used will be different and the way that the pattern of individual binary digits (bits) which is decoded into a number will be different.

number: the name or label for the variable. Once memory has been allocated to store our integer value, our programme can access that memory using the variable name.

=: This is an assignment operation that stores the number 12 in our variable.

When C++ declares a variable, a block of the computer's memory will be used to store the value, it may already contain a bit pattern (from operations performed elsewhere in your program or another program running on the computer). This means that when we create a new variable, it will contain a random value. Therefore, it is important that you assign a value to a variable before you start to use it.

The program is below(see Figure 3-1).

```
1  #include <iostream>
2  using namespace std;
3
4  int main()
5  {
6      int a = 3, b = 5;          // initializes a and b.
7      float c = 22.5;            // initializes c.
8      double pi = 3.14159;       // declares an approximation of pi.
9      char x = 'x';              // the variable x has the value 'x'.
10
11     return 0;
12 }
```

Figure 3-1   Assignment for different variables

3.5 Variable Names and Comments

Often in some C/C++ textbooks or when you look at coding examples on the web, you see code that looks like:

```
int a, b, c;
float d, e;
```

This is valid C/C++ code, the first line declares three variables of type int with the names or labels of a, b and c. The second line declares two variables to hold floating point data and gives them the names d and e.

Although this code will work, it is not recommended when writing C/C++ programs. It is a legacy of the times when computers were very limited in terms of memory storage and computing power. Today, the mobile phone in your pocket or a laptop computer on your desk have computing power more than a mainframe computer that would have been shared between 20 to 30 users at the year when C was developed.

Modern C/C++ programming practice will use longer names for variables. The name given to each variable should show what the variable will store.

If we have a variable called a, what does it store? Is it the number of apples in a shop or the number of pigs that a farmer owns or the telephone number of a friend? The variable name tells the person reading the program nothing useful. This means that we would have to add comments to our code. Comments are text in our source file that is ignored by our compiler.

程序注释

A single line comment is indicated by using the symbol "//". The compiler will ignore all characters after the "//" symbol until the end of the line. Our code with comments might look like:

单行注释

```
int a;    //used to store the number of apples in shop
int b;    //used to store the number of pigs on a farm
int c;    //used to store current telephone number
```

Even with comments we have to remember what each variable in the program does, and we may need to keep looking back at the comments so that we are reminded of what the variable holds. The preferred technique in modern coding is to use what is known as self-commenting. In this approach the name of a variable indicates what it is stored in the variable. Using this approach, we would declare our variables as:

技术

自我注释

```
int applesInShop;
int pigsOnFarm;
int currentPhoneNumber;
```

You can see, we hope, that the variable names show the information that we expect the variable to hold.

Styles of Variable Names

变量命名风格

C/C++ does not allow us to have space characters in a variable name. So we need to be able to tell where the individual words in the variable name start. Some people use underscore characters "_" where the spaces are:

```
apples in shop becomes apples_in_shop
```

In this book, we are using "camel case" where we don't use underscore

驼峰式命名方式

characters to represent spaces, but we use a capital or upper case letter at the start of each word in the variable name.

```
apples in shop becomes applesInShop
```

You can also see that the variable name starts with a small or lower case letter. We will always follow this style throughout this book and you should do the same.

Unlike some scripting languages, in C/C++ you must formally declare a variable, that is, to define its type and name before we can use it elsewhere in our program. This strict typing approach means that we have to think a little more when we are writing our code but are less likely to have our code failing to work correctly when it is executed.

脚本语言；正式声明一个变量

C/C++ allows you to declare variables without assigning values to them in which case the variable will have a random value, this can cause your program to perform incorrectly or crash. It is good practice to give a variable a value when you create it, and avoiding the circumstances such as dividing by zero, multiplying by infinity, etc.

赋值
随机值
崩溃

乘以一个无穷大的数

If you want to store a different value to your variable, then you need to assign a new value to the variable using the assignment operator "=", the general form for assignment is:

```
variableName = variableValue;
```

The value of a variable can be changed by setting a literal value, or by assigning another variable to it, in which case the variable just takes on the value of the variable on the right, that is to say, the original value of the variable is overridden.

文本（值）

被覆盖了

```
newBunnies = 15;
bunnyCounter = newBunnies;
```

3.6 Operators

> An operator is a symbol that tells the compiler to perform specific mathematical or logical manipulations.

C++ is rich in built-in operators and provides the following types of operators:

- Arithmetic operators
- Relational operators
- Logical operators
- Bitwise operators
- Assignment operators

This chapter will examine the arithmetic, relational, logical, bitwise, assignment operators one by one.

3.6.1 Arithmetic Operators

As a engineer, much of the data processing you are most likely to perform is going to be arithmetic based. Most programming languages are geared around arithmetic, with C/C++ being no exception.

There are following arithmetic operators supported by C/ C++ language in Table 3-2.

Table 3-2   Arithmetic operators (Assume variable A holds 10 and variable B holds 20)

Operator	Description	Example
+	Addition, adds two operands	A+B will give 30
−	Subtraction, subtracts second operand from the first	A−B will give −10
*	Multiplication, multiplies both operands	A*B will give 200
/	Division	B/A will give 2
%	Modulus operator, remainder of after an integer division	B%A will give 0
++	Increment operator, increases integer value by one	A++ will give 11
−−	Decrement operator, decreases integer value by one	A−− will give 9

We often find in the code that we want to add something to the contents of a variable and store the results in the original variable. The following line of code will perform this task.

```
applesInShop += applesFromFarm;
```

The operator "+=" takes the contents of the variable applesFromFarm and adds it to the contents of the variable applesInShop where the result of the operation is stored. It is equivalent to the statement below:

```
applesInShop = applesInShop +applesFromFarm;
```

There is a minor advantage in terms of the use of memory by using this operator. 小的优点(内存的占用小一些)

There are corresponding operators for subtraction, multiplication and division.

Table 3-3 Arithmetic operators

Operation	Original values		Value stored in variable **a** after operation
	a	b	
a += b;	4	2	6
a -= b;	4	2	2
a *= b;	4	2	8
a /= b;	4	2	2

3.6.2 Logical Operators

逻辑运算符

There are three logical operators supported by C/C++ language: logical AND, logical OR and logical NOT.

Table 3-4 gives a explaination for each logical operator.

Table 3-4 Logical operators (Assume variable A holds 1 and variable B holds 0)

Operator	Description	Example
&&	Logical AND. If both the operands are true, then condition becomes true	(A&&B) is false
||	Logical OR. If any of the two operands is true, then condition becomes true	(A||B) is true
!	Logical NOT. Use to reverses the logical state of its operand. If a condition is true, then Logical NOT operator will make false	! A is false ! B is true

3.6.3 Bitwise Operators

位运算符

A bitwise operator works on bits and perform bit-by-bit operation. The truth table for &, |, and ^ are as follows:

位与位操作

Table 3-5   Bitwise operators in C/C++

p	q	p & q	p | q	p ^ q
0	0	0	0	0
0	1	0	1	1
1	1	1	1	0
1	0	0	1	1

The Bitwise operators supported by C/C++ language are listed in Figure 3-2. Assume variable A holds 60 and variable B holds 13, then:

```cpp
#include <iostream>
using namespace std;

main()
{
    unsigned int a = 60;   // 60 = 0011 1100
    unsigned int b = 13;   // 13 = 0000 1101
    int c = 0;

    c = a & b;             // 12 = 0000 1100
    cout << "Line 1 - Value of c is : " << c << endl;
    c = a | b;             // 61 = 0011 1101
    cout << "Line 2 - Value of c is: " << c << endl;
    c = a ^ b;             // 49 = 0011 0001
    cout << "Line 3 - Value of c is: " << c << endl;
    c = ~a;                // -61 = 1100 0011
    cout << "Line 4 - Value of c is: " << c << endl;
    c = a << 2;            // 240 = 1111 0000
    cout << "Line 5 - Value of c is: " << c << endl;
    c = a >> 2;            // 15 = 0000 1111
    cout << "Line 6 - Value of c is: " << c << endl;
    return 0;
}
```

Figure 3-2   Bitwise operators example

The output of the program above is as Figure 3-3.

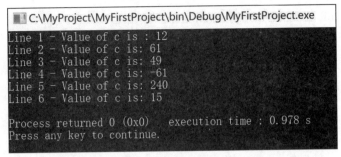

Figure 3-3 Output of the bitwise operators example

3.6.4 Relational Operators

There are following relational operators supported by C/C++ language in Table 3-6.

Table 3-6 Relational operators in C/C++ (Assume variable A holds 10 and variable B holds 20)

Operator	Description	Example
==	Checks if the values of two operands are equal or not, if yes, then condition becomes true	(A==B) is not true
!=	Checks if the values of two operands are equal or not, if values are not equal, then condition becomes true	(A!= B) is true
>	Checks if the value of left operand is greater than the value of right operand, if yes, then condition becomes true	(A>B) is not true
<	Checks if the value of left operand is less than the value of right operand, if yes, then condition becomes true	(A<B) is true
>=	Checks if the value of left operand is greater than or equal to the value of right operand, if yes, then condition becomes true	(A>=B) is not true
<=	Checks if the value of left operand is less than or equal to the value of right operand, if yes, then condition becomes true	(A<=B) is true

3.6.5 Operators Precedence in C/C++

Operator precedence determines the value of an expression. Certain operators have higher precedence than others. For example, the multiplication operator has higher precedence than the addition operator. For example:

```
x = 7 + 3 * 2;
```

Here, x is assigned 13, not 20, because operator * has higher precedence than.

In Table 3-7, operators with the highest precedence appear at the top, those with the lowest appear at the bottom. Within an expression, higher precedence operators will be evaluated first.

Table 3-7 Operators Precedence in C/C++

Category	Operator	Associativity	
Postfix	() [] -> . ++ - -	Left to right	后置（运算符）
Unary	+ - ! ~ ++ - - (type)* & sizeof()	Right to left	一元运算符
Multiplicative	* / %	Left to right	乘、除、求余
Additive	+ -	Left to right	加、减
Shift	<< >>	Left to right	移位
Relational	< <= > >=	Left to right	关系运算
Equality	== !=	Left to right	相等性
Bitwise AND	&	Left to right	位与
Bitwise XOR	^	Left to right	位异或
Bitwise OR	|	Left to right	位或
Logical AND	&&	Left to right	逻辑与
Logical OR	||	Left to right	逻辑或
Conditional	?:	Right to left	条件运算
Assignment	= += -= *= /= %= >>= <<= &= ^= |=	Right to left	赋值
Comma	,	Left to right	逗号运算符

Below (see Figure 3-4) is the example for demonstrating the precedence of operators.

```cpp
#include <iostream>
using namespace std;

main() {
    int a = 20;
    int b = 10;
    int c = 15;
    int d = 5;
    int e;

    e = (a+b)*c/d;        // (30*15)/5
    cout << "Value of (a+b)*c/d is:" << e << endl;
    e = ((a+b)*c)/d;      // (30*15)/5
    cout << "Value of ((a+b)*c)/d is:" << e << endl;
    e = (a+b)*(c/d);      // (30)*(15/5)
    cout << "Value of (a+b)*(c/d) is:" << e << endl;
    e = a+(b*c)/d;        // 20+(150/5)
    cout << "Value of a+(b*c)/d is:" << e << endl;
    return 0;
}
```

Figure 3-4　Example of the precedence of operators

The output is as below (see Figure 3-5):

```
Value of (a+b)*c/d is:90
Value of ((a+b)*c)/d is:90
Value of (a+b)*(c/d) is:90
Value of a+(b*c)/d is:50

Process returned 0 (0x0)    execution time : 0.488 s
Press any key to continue.
```

Figure 3-5　Output for code of precedence of operators

3.7　Some Much Used Operators in C / C++

3.7.1　Increment and Decrement Operator

自增和自减运算符

The increment operator ++ adds 1 to its operand, and the decrement operator-- subtracts 1 from its operand. Both the increment and decrement operators can either precede (prefix) or follow (postfix) the operand.

前置的；后置的

When an increment or decrement is used as part of an expression, there is an important difference in prefix and postfix forms.

> Increment or decrement will be done before rest of the expression for a prefix increment or decrement operation, and increment or decrement will be done after the complete expression is evaluated for a postfix increment or decrement operation.

Let's see some examples of ++ as prefix and postfix in C and C++ (see Figure 3-6).

```cpp
#include <iostream>
using namespace std;

int main()
{
    int var1 = 1, var2 = 1;

    cout << var1++ << endl; // var1 is displayed as 1, then it is increased to 2.
    cout << var1++ << endl;


    cout << ++var2 << endl; // var2 is increased to 2 then, it is displayed.

    return 0;
}
```

Figure 3-6　Increment and decrement operators

3.7.2　sizeof() Operator

The sizeof() operator is a compiling time unary operator which can be used to compute the size of its operand. The result of sizeof() is of unsigned integral type. sizeof can be applied to any data-type, including primitive types such as integer and floating-point types, pointer types, or compound datatypes such as classes, structures, unions and any other user defined data type. The syntax of using sizeof is as follows:

一元运算符
无符号的整数类型（的数）
指针类型的；混合类型，比如类、结构体、联合体（都是数据类型）

```
sizeof (data type)
```

The following examples demonstrate the sizeof operator available in C and C++ (see Figure 3-7).

```cpp
int main()
{
    cout << "Size of char : " << sizeof(char) << endl;
    cout << "Size of int : " << sizeof(int) << endl;
    cout << "Size of float : " << sizeof(float) << endl;
    cout << "Size of double : " << sizeof(double) << endl;
    cout << "Size of bool : " << sizeof(bool) << endl;

    return 0;
}
```

Figure 3-7　Demonstration of the sizeof operator

The output is:

```
Size of char : 1
```

```
Size of int : 4
Size of float : 4
Size of double : 8
Size of bool : 1
```

Note: sizeof() may give different output according to machines.

3.7.3 Modulus (%) Operator

模运算符（求余）

The modulus operator produces the remainder of an integer division.

余数

Syntax: If x and y are integers, then the expression:

```
x % y
```

produces the remainder when x is divided by y.

> The % operator cannot be applied to floating-point numbers, i.e float or double. If you try to use the modulus operator with floating-point constants or variables, the compiler will produce an error.

- If x is completely divided by y, the result of the expression is 0.
- If x is not completely divisible by y, then the result will be the remainder in the range [1, y−1].
- If y is 0, then a compile-time error will be produced.

整除

Basic code for explaination is as below (see Figure 3-8).

```
1  #include <stdio.h>
2  int main()
3  {
4      int x, y, z;
5      x = 6;
6      y = 5;
7      z = 3;
8
9      printf("%d\n", x % y);
10     printf("%d\n", y % x);
11     printf("%d\n", x % z);
12     printf("%d\n", z % x);
13
14     return 0;
15 }
```

Figure 3-8 Demonstration of modulus operator

3.7.4 Conditional Operator (? :)

条件运算符

```
variable = expression1 ? expression2 : expression3;
```

Where expression1, expression2, and expression3 are expressions, variable holds the value of the entire expression.

First, expression1 is evaluated, if it is true, then expression2 is evaluated and becomes the value of the entire expression. If expression1 is false, then expression3 is evaluated and its value becomes the value of the expression.

The conditional operator is a kind of similar to the if-else statement that we will learn in the next chapter. It follows the same algorithm as of if-else statement but the conditional operator takes less space and helps to write the if-else statements in the shortest way. It can be visualized into if-else statement as:

```
if(expression1==true)
{
    variable = expression2;
}
else
{
    variable = expression3;
}
```

Since the conditional operator takes three operands to work, hence it is also called ternary operator, and it is the only ternary operators in C and C++.

三元运算符

Below is the example of conditional operator.

```
#include<iostream>
using namespace std;
int main()
{
    int x = 3, y = 5, bigNumber;
    bigNumber = (x > y ) ? x : y;
    cout << "The big number is: " << bigNumber << endl;
    return 0;
}
```

3.7.5 comma "," operator

逗号运算符

In a C/C++ program, comma is used in two contexts: (1) A separator (2) An

Operator. For example:

```
#include<iostream>
using namespace std;
int main()
{
    int a = 3, b = 4, c = 5;    //commas are used as separators
    cout << "a=" << a << endl;
    cout << "b=" << b << endl;
    cout << "c=" << c << endl;
    return 0;
}
```

In the example above, commas work as separators, not as operators. The first line of statement declares three variables and then do assignment one by one from left to right. Below is the example for comma operator used as an operator:

```
#include<iostream>
using namespace std;
int main()
{
    int a;
    a = 1, 2, 3;
    cout<< a << endl;
    return 0;
}
```

In the program above, comma works as an operator. The comma operator has the lowest precedence of any operator, so the assignment operator takes precedence over comma and the expression "a = 1, 2, 3" becomes equivalent to "(a = 1), 2, 3". The output is 1.

最低的运算优先级

For the program segment below:

```
int a=1, b=2, c=3, d;
d = (a, b, c);
```

程序段

Commas act as separators in the first line and as an operator in the second line.

In the second line, the round brackets are used, so comma operator is executed first. The comma expression (a, b, c) is a sequence of expressions which

圆括号
顺序(执行)表达式

evaluates to the last variable c, so the value of d is 3.

Look at the program segments below:

```
int a=3, b=4, c=5, d;
d = (a += 1, a + b, a + c);
```

For the line 2 of statement, the expression (a += 1, a + b, a + c) is separated into three parts by two commas:

The first part increases value of a by 1, so a holds 4.

The value of the second part is 8 (4+4).

The value of the third part is 9 (4+5).

As the value of the entire expression (a += 1, a + b, a + c) is the value of third part, so the value of d is 9.

Chapter Review

1. What is a variable?

2. Evaluate the following expressions:

   a. 12*3 + 4

   b. 5+3/2

   c. 4/5*5

   d. 11%3

   e. (4+5)/3

Programming Exercises

1. Write a program in C to print the sum of two numbers, and then change this program into a C++ language program.

2. Write a program in C++ to find the size of fundamental data types using the sizeof() operator. The output should look like this:

```
Find Size of fundamental data types :
--------------------------------------------
The sizeof char is: 1 bytes.
The sizeof int is: 4 bytes.
The sizeof float is: 4 bytes.
The sizeof double is: 8 bytes.
The sizeof bool is: 1 bytes.
```

3. Assume an integer takes 4 bytes, what is the output of the program below?

```
#include<stdio.h>
int main()
{
    int i = 5, j = 10, k = 15;
    printf("%d ", sizeof(k /= i + j));
    printf("%d", k);
    return 0;
}
```

(A) 4 1

(B) 4 15

(C) 2 1

(D) Compile-time error

4. What is the output of the program below?

```
#include<stdio.h>
int main()
{
    int i = (1, 2, 3);
    printf("%d", i);
    return 0;
}
```

(A) 1

(B) 3

(C) Garbage value

(D) Compile time error

5. What is the output of the program below?

```c
#include<stdio.h>
int main()
{
    int a = 1;
    int b = 1;
    int c = a || --b;
    int d = a-- && --b;
    printf("a = %d, b = %d, c = %d, d = %d", a, b, c, d);
    return 0;
}
```

(A) a = 0, b = 1, c = 1, d = 0
(B) a = 0, b = 0, c = 1, d = 0
(C) a = 1, b = 1, c = 1, d = 1
(D) a = 0, b = 0, c = 0, d = 0

6. Write a program to produce the output as shown below:

```
Results:
x value      y value     expressions     results
10           | 5         | x+=y          | x=15
10           | 5         | x-=y-2        | x=7
10           | 5         | x*=y*5        | x=250
10           | 5         | x/=x/y        | x=5
10           | 5         | x%=y          | x=0
```

7. Write a program in C++ to find the area and perimeter of a rectangle.

Sample output is:

```
Find the Area and Perimeter of a Rectangle :
------------------------------------------
Please input the length of the rectangle : 3
Please input the width of the rectangle : 4
The area of the rectangle is : 12
The perimeter of the rectangle is : 14
```

8. Write a program in C++ to print the ASCII code of a given character.

   Sample output is:

```
Print the ASCII code of a given character:
---------------------------------------------------
Input a character: a
The ASCII value of a is: 97
The character for the ASCII value 97 is: a
```

Chapter 4   Decision Making

So far in the programmes that we have learned the code has executed sequentially, that means the execution started with the first line of the main() function and each line in main() is executed in turn. This means that there is only one path through our programme.

In the real world, this would be very limiting. Think about what happens with a common every day activity like using an ATM (electronic cash dispenser) at the bank.

It is very easy to describe to a human being how to use the ATM, you might say something like this:

"*You put your bank card into the machine and it will ask for your code number. You get three attempts to enter your code. If you get it wrong three times in a row, the machine will keep your card and you won't be able to get your money. If you get your code right, you will be able to use the machine to access your bank account and when you have finished, you will get your card back.*"

When you try to translate this behaviour to software, you need to use some features of C/C++ that haven't been looked at yet. The behaviour of the software is now conditional: it depends on whether the user has input the correct code number and also on the internal state of the programme which tracks how many attempts have been made to input the code correctly.

At the end of attempting to use the machine, there are two possible states:

Successful you have put the right code in within 3 attempts.

Unsuccessful you failed to enter the correct code three times.

There are 4 different paths that lead from the starting points to these two end states:

Chapter 4　Decision Making

- Code input correctly on the first attempt: go to successful end state;
- Code input wrongly on the first attempt but correctly on the second attempt: go to successful end state;
- Code input wrongly on the first and second attempts but correctly on the third attempt: go to successful end state;
- Code input wrongly on all three attempts go to unsuccessful end state.

We could show the paths through this code by using a flowchart as is shown in Figure 4-1.　流程图

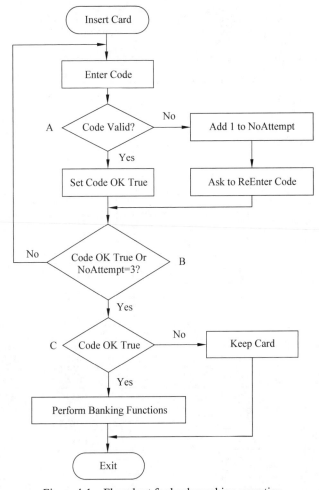

Figure 4-1　Flowchart for bank machine operation

Two of the decision boxes in the flow chart A and C are associated with what is　流程图

53

known as an "if else" conditional structure. In this structure, two possible branches of code can be followed: the first if the testing condition is true and the second if the condition is false.

The third decision box labelled B in figure 4-1 is associated with a LOOPING structure. A LOOP causes the program to repeat a section of code until a particular logical condition is fulfilled.

4.1 Conditional Operations

Before decisions are made in an IF ELSE and LOOPING, we need to be able to evaluate the condition that controls the path programme execution is going to follow. This evaluation requires operations that use variables within our programme as operands and return a value of type bool. We discussed the bool type in chapter 3 briefly and you should remember that it is a binary type which can only hold two possible values True or False. These operators are known as conditional operators (see Table 4-1).

Table 4-1 Conditional operators available in C/C++

Operator	Function	Example			
		a	b	Operation	Result
==	equality	2	2	a == b	True
		3	−2	a == b	False
!=	inequality	2	2	a != b	False
		3	5	a != b	True
>	greater than	4	2	a > b	True
		2	2	a > b	False
<	less than	4	2	a < b	False
		−2	0	a < b	False
>=	greater than or equal to	3	2	a >= b	True
		3	3	a >= b	True
<=	less than or equal to	3	2	a <= b	True
		3	3	a <= b	True

These conditional operators work for variables that contain numeric values, such as int and float. It is not good practice to use the equality operator "==" when working with floating point variables. This is because of errors that can result in the precision associated with the way that floating point numbers are represented by C/C++ variables. It is better to design your code so you can use the greater than or equals to >= operator or the less than or equal to <= operator.

Conditional operators and string objects

You can use the equality operator "==" and the inequality operator "!=" with string objects defined in the <string> library. The equality operator returns true only if the characters stored in the two strings are identical. The inequality operator will return true if there is any difference between the characters stored in the two string objects.

4.2 The if Structure

Sometimes, in our code, we want certain lines of our programme only to be executed if a variable or a set of variables hold particular values. We can achieve this using an if structure. The flow chart for the if structure is shown in Figure 4-2.

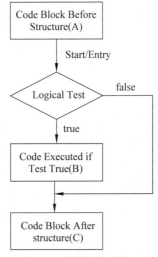

Figure 4-2   if structure flowchart

The if structure starts with the logical test block. It ends where the two execution paths associated with the true and false conditions merge. Note that although there is more than one path for our program execution within the structure there is only one exit point. This is an important feature of all structures. They contain multiple paths for programme execution but have a single entry point and a single exit point.

融合

出口
单入口
单出口

In the flowchart, there are two code blocks that are always executed: Block (A) before the structure and Block (C) after the structure. The code block Block (B) is only executed if the logical test is evaluated as true.

流程图

4.2.1　The if Statement

if 语句

An if statement conditionally executes one or more statements based on whether a specified expression is true.

有条件地执行

The if keyword is used to execute a statement or block, based on whether a specified expression is true. Its syntax is:

```
if (condition)
    statement;
```

Here, condition is the expression that is being evaluated. If this condition is true, statement is executed. If it is false, statement is not executed (it is simply ignored), and the program continues right after the entire selection statement.

For example, the following code fragment (see Figure 4-3) prints the message (x is positive) only if the value stored in the x variable is bigger than 0:

```
 1  #include<iostream>
 2  using namespace std;
 3
 4  main()
 5  {
 6      int a=5,b=3,x;
 7      x=a-b;
 8      if(x>0)
 9          cout << "x is positive" << endl;
10
11      return 0;
12  }
```

Figure 4-3　Code example for one way selection

If x is not bigger than 0, code of line 9 will be is ignored, and nothing is printed.

If you want to include more than one single statement to be executed when the condition is fulfilled, these statements shall be enclosed with curly braces {}, forming a compound statement. For example (see Figure 4-4):

满足
复合语句

```
1   #include<iostream>
2   using namespace std;
3
4   main()
5   {
6       int a=5,b=3,x;
7       x=a-b;
8       if(x>0)
9       {
10          cout << "x is positive" << endl;
11          cout << "x = " << x << endl;
12      }
13
14      return 0;
15  }
```

Figure 4-4　Code example for compound statement

The output should be (see Figure 4-5):

```
C:\MyProject\MyFirstProject\bin\Debug\MyFirstProject.exe        —    □    ×
x is positive
x = 2

Process returned 0 (0x0)    execution time : 0.558 s
Press any key to continue.
```

Figure 4-5　The output of the program in Figure 4-4

The condition can be of any type that can be converted to bool, which means it can be type of int, float, double, or even char. These variables must be initialized and converted to bool. Please bear in mind:

> Zero is converted to logical false, none zero numbers are converted to true.

4.2.2　The if-else Statement

if-else 语句

Selection statements with if can also specify what happens when the condition is not fulfilled, by using the else keyword to introduce an alternative statement. Its syntax is:

```
if (condition)
    statement1;
else
    statement2;
```

Where statement1 is executed in case condition is true, and in case it is not, statement2 will be executed. For example (see Figure 4-6):

```
1   #include<iostream>
2   using namespace std;
3
4   main()
5   {
6       int a=5,b=3,x;
7       x=a-b;
8       if(x>0)
9           cout << "x is positive" << endl;
10      else
11          cout << "x could be zero or negative" << endl;
12
13      return 0;
14  }
```

Figure 4-6   Code example for if-else statement

4.2.3 Nested if Statements

嵌套的 if 语句

The if-else statement can be more complex. You can use one if-else statement inside another if or else statement(s) (see Figure 4-7). These are known as nested if statements.

```
1   #include<iostream>
2   using namespace std;
3
4   main()
5   {
6       int a=5,b=3,x;
7       x=a-b;
8       if(x>0)
9           cout << "x is positive" << endl;
10      else
11          if(x==0)
12              cout << "x is zero" << endl;
13          else
14              cout << "x is negative" << endl;
15
16      return 0;
17  }
```

Figure 4-7   Code example for nested if statements

Remember, whitespace and indentation are a convenience for the programmer;

they make no difference to the compiler.

4.2.4 if-else-if Statement

if-else-if statement is used when you need to check multiple conditions. In this control structure, you have only one "if" at the beginning and one "else" at the end, you can have multiple "else if" blocks in the middle. This is how it looks (see Figure 4-8):

多个条件

```
if(condition_1)
{
    statement(s);  ─── execute this block if condition_1 is true
}
else if(condition_2)
{
    statement(s);  ─── execute this block if condition_1 is not true
                       and condition_2 is true
}
else if(condition_3)
{
    statement(s);  ─── execute this block if condition_1 and condition_2
                       are not true and condition_3 is true
}
⋮
else
{
    statement(s);  ─── if none of the conditions above are true
                       then these statements gets executed
}
```

Figure 4-8 The structure of if-else-if statement

For example, a student's grade point average (GPA) and scores can be converted by the Table 4-2. Write a program to prompt a user to enter a student's score, then output the GPA (see Figure 4-9).

Table 4-2 Score and GPA

Score	GPA
90≤score≤100	4.0
85≤score<90	3.7
82≤score<85	3.3
78≤score<82	3.0
75≤score<78	2.7
72≤score<75	2.3
68≤score<72	2.0
64≤score<68	1.7
60≤score<64	1.0
Score<60	0

```cpp
#include<iostream>
using namespace std;

int main()
{
    float score;
    float gpa;
    cout << "Please enter the score: ";
    cin >> score;
    if(score>100)
    {
        cout << "Input Error!" << endl;
        exit(0);
    }
    else if(score>=90)
    {
        gpa = 4.0;
    }
    else if(score>=85)
    {
        gpa = 3.7;
    }
    else if(score>=82)
    {
        gpa = 3.3;
    }
    else if(score>=78)
    {
        gpa = 3.0;
    }
    else if(score>=75)
    {
        gpa = 2.7;
    }
    else if(score>=72)
    {
        gpa = 2.3;
    }
    else if(score>=68)
    {
        gpa = 2.0;
    }
    else if(score>=64)
    {
        gpa = 1.7;
    }
    else if(score>=60)
    {
        gpa = 1.0;
    }
    else
    {
        gpa = 0.0;
    }
    cout << "GPA is " << gpa << "." <<endl;

    return 0;
}
```

Figure 4-9　Code for if-else-if statement

4.2.5 Switch-case statement

Switch-case statement is used when we have multiple conditions and we need to perform different action based on the conditions. When we have multiple conditions and we need to execute a block of statements when a particular condition is satisfied, the switch-case statement is an efficient way to handle such scenarios, easy to understand and easy to use.   场景

The syntax for a switch-case statement in C/C++ is as follows (see Figure 4-10):

```
switch(expression)
{
  case constant-expression 1:
     statement(s);
     break; //optional
  case constant-expression 2:
     statement(s);
     break; //optional
  
  // you can have any number of case statements.
  default: //Optional
     statement(s);
}
```

Figure 4-10 The structure of switch statement

The following rules apply to a switch-case statement (see Figure 4-11):

- Each case is followed by the value to be compared to and a colon.
- The constant-expression for a case must be the same data type as the variable in the switch, and it must be a constant or a literal.   数值型常量或字符型常量
- When the variable being switched on is equal to a case, the statements following that case will execute until a break statement is reached.
- When a break statement is reached, the switch terminates, and the flow of control jumps to the next line following the switch statement.
- Not every case needs to contain a break. If no break appears, the flow of control will fall through to subsequent cases until a break is reached.
- A switch statement can have an optional default case, which must appear at the end of the switch. The default case can be used for performing a task when none of the cases is true. No break is needed in the default case.

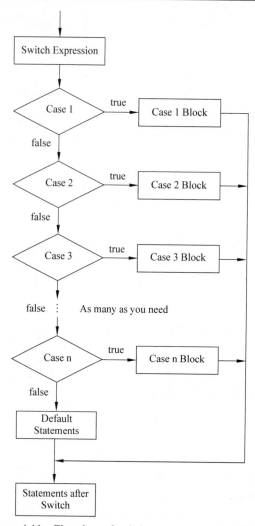

Figure 4-11   Flowchart of switch-case statement in C/C++

The example (Figure 4-12) below uses the weekday number to calculate the weekday name.

Try to remove the break statements, then run the program to see what happened.

```cpp
1  #include<iostream>
2  using namespace std;
3  int main()
4  {
5      int day;
6      cout << "Please enter a digit(1-7) :";
7      cin >> day;
8      switch (day)
9      {
10         case 1:
11             cout << "Monday";break;
12         case 2:
13             cout << "Tuesday";break;
14         case 3:
15             cout << "Wednesday";break;
16         case 4:
17             cout << "Thursday";break;
18         case 5:
19             cout << "Friday";break;
20         case 6:
21             cout << "Saturday";break;
22         case 7:
23             cout << "Sunday";break;
24         default:
25             cout << "Input Error!";
26     }
27     return 0;
28 }
```

Figure 4-12   Code example for switch statement

Figure 4-13 shows the difference of switch statement between C and C++.

```cpp
#include<iostream>
using namespace std;
int main()
{
    char ch;
    cout << "Please input a,b or c: ";
    cin >> ch;
    switch(ch)
    {
        case 'a':
            cout<<"a is entered.";
            break;
        case 'b':
            cout<<"b is entered.";
            break;
        case 'c':
            cout<<"c is entered.";
            break;
        default: cout<<"Input error!";
    }
    return 0;
}
```

```c
#include<stdio.h>
int main()
{
    char ch;
    printf("Please input a,b or c:");
    scanf("%c",&ch);
    switch(ch)
    {
        case 'a':
            printf("a is entered.");
            break;
        case 'b':
            printf("b is entered.");
            break;
        case 'c':
            printf("c is entered.");
            break;
        default:
            printf("Input error!");
    }
    return 0;
}
```

Figure 4-13   Code example written in C and C++

Programming Exercises

1. Write a C++ program that prompts the user to input three integer values and find the greatest value.

2. Write a program that determines a student's grade. The program will read three types of scores (quiz, mid-term, and final scores) and determine the grade based on the following rules:
   -if the average score> =90: grade=A
   -if the average score >= 70 and <90: grade=B
   -if the average score>=60 and <70 : grade=C
   -if the average score<60 : grade=D

3. Write a C++ program to compute the real roots of the equation: $ax^2+bx+c=0$.   实根
   The program will prompt the user to input the values of a, b, and c. It then computes the real roots of the equation based on the following rules:
   -if a and b are zero=> no solution
   -if a is zero=>one root (−c/b)
   -if b^2-4ac is negative=>no roots
   -Otherwise=> two roots
   The roots can be computed using the following formula:
   $x_1 = -b+(b^2-4ac)^{1/2}/2a$
   $x_2 = -b-(b^2-4ac)^{1/2}/2a$

4. Find the absolute value of a number entered by the user.   绝对值

5. Write a program to check whether a triangle is valid or not.   三角形

6. The marks obtained by a student in 5 different subjects are input by the user.
   The student gets a division as the following rules: 等级
   Percentage above or equal to 60 - First division   甲等
   Percentage between 50 and 59 - Second division   乙等
   Percentage between 40 and 49 - Third division   丙等
   Percentage less than 40 – Fourth division   丁等
   Write a program to calculate the division obtained by the student.

7. Write a program to check if a given year is leap year or not.
   A year is leap year if the following conditions are satisfied:
   - Year is exactly divisible by of 400.
   - Year is exactly divisible by 4 and not divisible by 100.

8. Write a program to determine whether the character entered is a capital letter, a small case letter, a digit or a special symbol. Table 4-3 shows the range of ASCII values for various characters.

Table 4-3   The   range of ASCII values for various character

Characters	ASCII Values
A~Z	65 ~ 90
a ~ z	97 ~ 122
0 ~ 9	48 ~ 57
special symbols	0 ~ 47, 58 ~ 64, 91 ~ 96, 123 ~ 127

Chapter 5   Loops

In general, statements are executed sequentially: The first statement in a function is executed firstly, followed by the second, and so on. However, there may be a situation, when you need to execute a block of code many times.

顺序执行

A loop statement allows us to execute a statement or group of statements multiple times until a particular condition is satisfied.

循环语句

For example, let's say we want to show a message 10 times. Instead of writing the print statement 10 times, we can use a loop. That is just a simple example. We can achieve much more efficiency and sophistication in our programs by making effective use of loops.

精妙的作用

There are 3 types of loops in C/C++:

- for loop;
- while loop;
- do...while loop.

for 循环
while 循环
do...while 循环

5.1   for Loop

The syntax of for-loop is:

```
for (initialization; condition; update)
{
    // body of-loop
}
```

Here,

initialization—initializes variables and will be executed only once;

初始化；给循环变量赋初值

condition—if true, the body of for loop is executed;

if false, the for loop is terminated;

终止

update—updates the value of initialized variables.   更新循环变量的值

Figure 5-1 is the flowchart of a for-loop in C/C++.

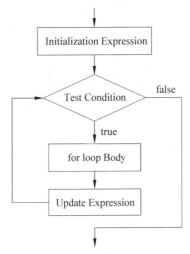

Figure 5-1   Flowchart of a for-loop in C/C++

Example: Printing Numbers From 1 to 5, the code is as below.

```
// C++ program: printing numbers from 1 to 5
#include<iostream>
using namespace std;
int main()
{
   int i=1;
   for(i=1; i<=5; i++)
      cout << i << " ";

   return 0;
}
```

The following program is written in C.

```
// C program: printing numbers from 1 to 5
#include<stdio.h>
int main()
{
   int i=1;
   for(i=1; i<=5; i++)
```

```
        printf("%d ", i);

    return 0;
}
```

The output is:

```
1 2 3 4 5
```

How does this program work?

Iteration	Value of i	i <= 5	Action
1st	i=1	true	1 printed, i is increased to 2
2nd	i=2	true	2 printed, i is increased to 3
3rd	i=3	true	3 printed, i is increased to 4
4th	i=4	true	4 printed, i is increased to 5
5th	i=5	true	5 printed, i is increased to 6
6th	i=6	false	Loop is terminated

Example: Find the sum of: 1+2+3+···+100.

```
int main()
{
    int i, sum=0;
    for(i=1; i<=100; i++)
    {
        sum = sum + i;
    }
    cout << "Sum = " << sum;

    return 0;
}
```

In the above example, we have two variables i and sum. The sum variable is initialized to 0, and used to hold the sum of the formula. 初始化为

Let us look at the for loop in more detail:

```
for (i = 1; i <= 100; i++)
```

Here,

i = 1: initializes the count variable;

i <= 100: runs the loop as long as i is less than or equal to 100;

i++: increase the i variable by 1 in each iteration.

When i becomes 101, the condition is false and sum will be equal to 1+2+3+⋯+100.

5.2   while Loop

The syntax of the while loop is:

```
while(condition)
{
    // body of the loop
}
```

Here is how a while loop works:

If the condition evaluates to true, the code inside the while loop is executed.

Then the condition is evaluated again.

This process continues until the condition is false.

When the condition evaluates to false, the loop terminates.

Figure 5-2 is the flowchart of a while-Loop in C/C++.

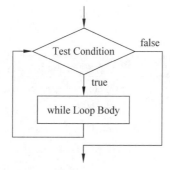

Figure 5-2   Flowchart of a while-Loop in C/C++

Example: Display numbers from 1 to 5.

```cpp
// C++ Program to print numbers from 1 to 5
#include<iostream>
using namespace std;
int main()
{
    int i = 1;
    while (i <= 5)
    {
        cout << i << " ";
        ++i;
    }

    return 0;
}
```

Display numbers from 1 to 5 in C:

```c
// C Program to print numbers from 1 to 5
#include<stdio.h>
int main()
{
    int i = 1;
    while (i <= 5)
    {
        printf("%d ",i);
        ++i;
    }

    return 0;
}
```

The output of the programs is as below.

Iteration	Value of i	i <= 5	Action
1st	i=1	true	1 printed, i is increased to 2
2nd	i=2	true	2 printed, i is increased to 3
3rd	i=3	true	3 printed, i is increased to 4
4th	i=4	true	4 printed, i is increased to 5
5th	i=5	true	5 printed, i is increased to 6
6th	i=6	false	Loop is terminated

5.3 do...while Loop

The do...while loop is a variant of the while loop with one important difference: the body of do...while loop is executed once before the condition is checked. 变形

Its syntax is:

```
do
{
   // body of loop;
}
while(condition);
```

Here, The body of the loop is executed at first. Then the condition is evaluated.

If the condition evaluates to true, the body of the loop inside the do statement is executed again.

The condition is evaluated once again.

If the condition evaluates to true, the body of the loop inside the do statement is executed again.

This process continues until the condition evaluates to false. Then the loop stops.

Figure 5-3 is the flowchart of a do…while loop.

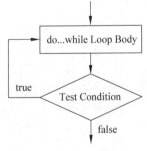

Figure 5-3   Flowchart of a do…while loop

Example: Display numbers from 1 to 5 using do…while loop.

```
#include<iostream>
using namespace std;
```

```
int main()
{
    int i = 1;
    do
    {
        cout << i << " ";
        ++i;
    }
    while (i <= 5);

    return 0;
}
```

The output is as below:

Iteration	Value of i	i <= 5	Action
	i=1	true	1 printed, i is increased to 2
1st	i=2	true	2 printed, i is increased to 3
2nd	i=3	true	3 printed, i is increased to 4
3rd	i=4	true	4 printed, i is increased to 5
4th	i=5	true	5 printed, i is increased to 6
5th	i=6	false	Loop is terminated

5.4   for Loops vs while Loops

A for loop is usually used when the number of iterations is known. For example,

```
// This loop is iterated 5 times
for (int i = 1; i <= 5; ++i)
{
    // body of the loop
}
```

Here, we know that the for-loop will be executed 5 times.

However, while and do...while loops are usually used when the number of iterations is unknown. For example,

```
while (condition)
{
    // body of the loop
}
```

5.5 Loop Control Statement "break" and "continue"

"break" and "continue" are two loop control statements in C and C++. They are only consisted of one keywork: break and continue.

循环控制语句

5.5.1 "Break" Statement

The break statement is a loop control statement which is used to terminate the loop.

终止循环

As soon as the break statement is encountered from within a loop, the loop iterations stop there and control jumps to the first statement after the loop immediately.

The syntax of the break statement is simple:

```
break;
```

Example : break with for loop

```
// program to print the value of i
#include<iostream>
using namespace std;
int main()
{
    for (int i = 1; i <= 5; i++)
    {
        // break condition
        if (i == 3)
        {
            break;
        }
        cout << i << endl;
    }
```

```
    return 0;
}
```

In the above program, the for loop is used to print the value of i in each iteration. Please notice the code:

```
if(i == 3)
{
    break;
}
```

This means, when i is equal to 3, the break statement executes and the loop is terminated. Hence, the output doesn't include values greater than or equal to 3.

The code below demonstrates the use of break statement in a while loop.

```
// Example: break with while loop
// program to find the sum of positive numbers
// if the user enters a negative number, break ends the loop
// the negative number entered is not added to sum

#include<iostream>
using namespace std;
int main()
{
    int number, sum = 0;
    while(true)
    {
        cout << "Enter a number: ";   //take input from the user
        cin >> number;
        if (number < 0)               //break condition
            break;
        sum += number;                //add all positive numbers
    }
    cout << "The sum is " << sum << endl; //display the sum
    return 0;
}
```

The code below shows the use of break in a nested for loop.

```
// C++ program to illustrate using break statement in nested loops
```

```cpp
#include<iostream>
using namespace std;
int main()
{
    // nested for loops with break statement at inner loop
    for (int i = 0; i < 5; i++)      //outter loop
    {
        for (int j = 0; j < 5; j++)  //inner loop
        {
            if (j > 3)
                break;
            else
                cout << "*";
        }
        cout << endl;
    }
    return 0;
}
```

外层循环

内层循环

In the code above, we can clearly see that the inner loop is programmed to execute for 5 iterations. But when the value of j becomes greater than 3, the inner loop stops executing which restricts the number of iteration of the inner loop to 4 iterations only (j from 0 to 3). However, the iteration of outer loop is unaffected.

限制

未受影响

When the program runs, we can see on the screen:

```
****
****
****
****
****
```

5.5.2 "continue" Statement

continue 语句

Continue is also a loop control statement which is opposite to that of break statement, it terminates the current iteration, but executes the next iteration of the loop.

As the name suggests, the continue statement forces the loop to continue or execute the next iteration. When the continue statement is executed in the loop, the code inside the loop following the continue statement will be skipped and next iteration of the loop will begin.

跳过

For example: Consider the situation when you need to write a program which prints number from 1 to 10 and but not 4. It is specified that you have to do this using only one loop.

指定

Here comes the usage of continue statement. We can run a loop from 1 to 10 and every time we have to compare the value of iterator with 4. If it is equal to 4, we will use the continue statement to continue to next iteration without printing anything. Otherwise, we will print the value.

Below is the implementation of the above idea:

实现

```
int main()
{
    for (int i = 0; i < 10; i++)
    {
        // If i is equals to 4, skip it and continue to next iteration
        //without printing.
        if (i == 4)
            continue;
        else
            cout << i << " ";   //otherwise print the value of i
    }
    return 0;
}
```

The output is:

1 2 3 5 6 7 8 9 10

Chapter Review

1. Why is break useful in switch statements?

2. What is the functionality of a continue statement?

3. How many types of loop statements?

Programming Exercises

1. Given a number, print a triangular pattern, only one loop is allowed to use.

```
Please input a number: 6
Output:
*
* *
* * *
* * * *
* * * * *
* * * * * *
```

2. Write a program to print the following pattern.

```
1******
12*****
123****
1234***
12345**
123456*
1234567
```

3. "One hundred copper coins to buy one hundred chickens" is a famous mathematic problem in ancient China which the ancient Chinese mathematician Zhang Chujian put forward in his book "*Suan Jing*", the problem is described as follows:
5 copper coins can buy 1 rooster,
3 copper coins can buy 1 hen,
1 copper coin can buy 3 chicks.
Now, buy 100 chickens for 100 copper coins. How many roosters, hens and chicks are there?
To solve this problem in a mathematical way, you can abstract the problem into a formula group. Set the number of roosters to x, set the number of hens to y, set the number of chicks to z, then we can get the following equation group:

百钱百鸡问题

《算经》

铜钱

$$\begin{cases} 5x+3y+1/3z = 100 \\ x+y+z = 100 \\ 0 <= x <= 20 \\ 0 <= y <= 33 \\ 0 <= z <= 100 \end{cases}$$

So this problem can be attributed to finding the integer solution of this indefinite equation. 不定方程

Solving indefinite equations by programming is different from manual calculations. Under the premise of analyzing and determining the range of the unknowns in the equation, the exhaustive range of the unknowns can be used to verify the conditions under which the equations are established and the corresponding solutions can be obtained.

人工的
前提，假设
未知数；穷尽的
验证

Extension: In the realization of this type of indefinite equation, the control variables of each layer of the loop are directly related to the unknown numbers of the equation, and the value range of the unknown numbers are used to cover all possible solutions. Based on the meaning of the problem, more reasonable setting of loop control conditions need to be considered, in order to reduce the number of such exhaustion and combination, improve the efficiency of the execution of the program.

控制变量
每层循环

覆盖所有可能的情况

4. Find out all the prime numbers from 2~100. A prime is a natural number greater than 1 that has no positive divisors other than 1 and itself. Examples of first few prime numbers are: 2, 3, 5,7,11,13.

素数

5. Find out all the narcissistic numbers.

水仙花数

A narcissistic number (also known as a pluperfect digital invariant) is a number that is the sum of its own digits each raised to the three power of the number of digits.

For example: 153 is a "narcissistic number" because $153 = 1^3 + 5^3 + 3^3$.

数字（个位数数字、十位数数字、百位数数字）

6. Write a program to calculate the sum of following series where n is input by user.

数列

$$1 + 1/2 + 1/3 + 1/4 + 1/5 +\cdots+1/n$$

7. Compute the sum of the following series

$$1 - 1/2 + 1/3 - 1/4 + 1/5 - \cdots + 1/n$$

where n is a positive integer and input by user.

8. Write a program to compute sin x for given x. The user should supply x and a positive integer n. We compute the sine of x using the series and the computation should use all terms in the series up through the term involving x^n

$$\sin x = x - x^3/3! + x^5/5! - x^7/7! + x^9/9! \cdots$$

9. Write a program to compute the cosine of x. The user should supply x and a positive integer n. We compute the cosine of x using the series and the computation should use all terms in the series up through the term involving x^n

$$\cos x = 1 - x^2/2! + x^4/4! - x^6/6! \cdots$$

Part 2　Core Language Features

Chapter 6　Arrays

Arrays are a vital part of programming especially if you are doing arithmetic operations or any form of numeric processing, which as engineers, you probably will be.

> An array is a series of elements of the same type placed in contiguous memory locations that can be individually referenced by adding an index to a unique identifier.

数组；至关重要的

元素
内存空间
特定的标识符
（指数组名）

That means that, for example, ten values of type float can be declared as an array without having to declare 10 different variables (each with its own identifier). Instead, using an array, the ten values are stored in contiguous memory locations, and all ten can be accessed using the same identifier, with the proper index.

连续的

索引

For example, an array containing 10 floating point values called a could be represented as Figure 6-1.

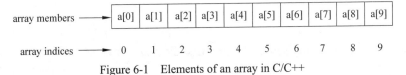

Figure 6-1　Elements of an array in C/C++

Where each panel represents one element of the array. In this case, these are values of type float. These elements are numbered from 0 to 9. In C or C++, the first element in an array is always numbered with zero (not a one), no matter its length.

6.1 Declaring an Array

Like a regular variable, an array must be declared before it is used. A typical declaration for an array is:

```
dataType arrayName[arraySize];
```

Here,

dataType——type of elements to be stored, such as int, float...

arrayName——name of the array, it should be a valid identifier.

arraySize——size of the array, which is always enclosed in square brackets ([]), specifies the length of the array or the number of elements. It must be an integer constant greater than zero.

For example, suppose we have 50 students in our class, and we need to store the scores of all of them. Instead of creating 50 separate variables, we can simply create an array:

```
float score[50];
```

The elements field within square brackets, representing the number of elements in the array, must be a constant expression, since arrays are blocks of static memory whose size must be determined at compiling time, before the program runs.

Things to Remember:

- The array indices start with 0. Meaning score[0] is the first element stored at index 0.
- If the size of an array is n, the last element is stored at index (n−1). In this example, score[49] is the last element.
- Elements of an array have consecutive addresses. Usually we do not need to know what the addresses are, but we do need to know that the addresses are consecutive.

The address of each element is increased by sizeof (float).

If the type of array is int, and for the most compliers the size of int is four bytes, then the address of each element is increased by 4.

If the type of array is char, and for the most compliers the size of char is one byte, then the address of each element is increased by 1.

6.2　Initializing an Array

Just like a normal variable, an array must be initialized before it can be referenced. There are two ways to initialize an array.　　　　　　　　　引用，使用

1. Initialize all the element during declaration:

```
// declare and initialize array mark
float mark[5] = {78,98,76,85,93};
```

2. Initialize each element one by one, there are two steps to do this.

　　First, declare a array, then initialize each element one by one:

```
float mark[5];
float mark[0] = 78;
float mark[1] = 98;
float mark[2] = 76;
float mark[3] = 85;
float mark[4] = 93;
```

or input the value of each element from the keyboard by a loop statement, the program is as below.

```
for(int i=0;i<=4;i++)
   cin >> mark[i];
```

6.3　Accessing Array Elements　　　　　　　　　　　　访问数组元素

An element is accessed by indexing the array name. This is done by placing the index of the element within square brackets after the name of the array.

Example: display sum and average of array elements using for loop, the program is as below.

```cpp
#include<iostream>
using namespace std;
int main()
{
    // initialization of the array.
    double numbers[6] = {97, 75, 82, 73, 95, 87};
    double sum = 0, count = 0, average;

    cout << "The numbers are: ";
    int i;
    for (i=0; i<6; i++)
    {
        cout << numbers[i] << " "; // print array elements in one line
        sum += numbers[i];         // calculate the sum
        ++count; i++;              // count the no. of array elements
    }
    cout << "\n Sum = " << sum << endl;      // print the sum
    average = sum / count;                    // find the average
    cout << " Average = " << average << endl;

    return 0;
}
```

The output is:

```
The numbers are: 97 75 82 73 95 87
Sum = 509
Average = 84.8333
```

In this program, a double array named numbers has been initialized. Three double variables sum, count and average are also been declared.

Here, sum =0 and count = 0.

Then we used a range based for loop to print the array elements. In each iteration of the loop, we add the current array element to sum.

基于（确定）范围的

We also increase the value of count by 1 in each iteration, so that we can get the size of the array by the end of the for loop.

After printing all the elements, we print the sum and the average of all the numbers. The average of the numbers is given by average = sum / count.

As we mentioned before, the elements are stored at contiguous memory locations. We can observe this through the example below:

```
// C++ program to demonstrate that array elements
// are stored contiguous locations
#include<iostream>
using namespace std;
int main()
{
    int a[5], i;
    cout << "Size of integer in this compiler is "
        << sizeof(int) << "\n";

    for (i = 0; i < 5; i++)
        cout << "Address a [" << i << "] is " << &a [i] << "\n";

    return 0;
}
```

The output will be:

```
Size of integer in this compiler is 4
Address a[0] is 0x61fe30
Address a[1] is 0x61fe34
Address a[2] is 0x61fe38
Address a[3] is 0x61fe3c
Address a[4] is 0x61fe40
```

In the example above, If a[0] is stored at address x, then a[1] is stored at x + sizeof(int), a[2] is stored at x + sizeof(int) *2, and so on.

"&" is address-of operator, use of "&" before a variable name, yields address of variable. 地址运算符；构成

6.4 Array out of Bounds

数组下标超界

If an array size is 10, then the array will contain elements from index 0 to 9.

If access to the element at index is 10 or more than 10, then the array index is out of bounds. However, the compiler will not produce an error, and the program can be built and run as normally. But, result may unpredictable and it will start causing many problems that will be hard to find. Therefore, you must be careful while using array indexing.

Let suppose an array have 5 elements, then the array indexing will be from 0 to 4. But, if an element with index is greater than 4 is accessed, result will be not you expected. For example:

```
//Program to demonstrate accessing array out of bounds
int main()
{
    int a[5] = {1,2,3,4,5};
    for(int i = 0; i <= 5; i++)
            cout << a[i] << " ";
    cout << endl;
    return 0;
}
```

In Microsoft Visual C++ 6.0, the output will be:

```
1 2 3 4 5 1703803
```

In CodeBlocks, the output will be:

```
1 2 3 4 5 0
```

In the example above, the programs are compiled and executed successfully. But while printing the element of a [5], the value of a[5] is unpredictable.

6.5   Address-of Operator (&)

The address of a variable can be obtained by preceding the name of the variable with an ampersand sign (&), known as address-of operator. For example:

```
pa = &a;
```

This would assign the address of variable a to pa. By preceding the name of the variable a with the address-of operator (&), we are no longer assigning the content of the variable itself, but its address. For example:

```
int main()
{
    // declare variables
    int a = 3;
    int b = 5;
    // print address of a
    cout << "Address of a: "<< &a << endl;
    // print address of b
    cout << "Address of b: " << &b << endl;

    return 0;
}
```

The output is:

```
Address of a: 0x61fe4c
Address of b: 0x61fe48
```

Here, 0x at the beginning represents the address is in hexadecimal form.

十六进制形式

Please note: The actual address of a variable in memory cannot be known before runtime, and the results may be different when you run the program.

实际的地址

运行时

6.6 Pointers in C and C++

指针

Whenever you create a variable in your code, for any type of variable, there are 4 pieces of information associated with it:

- The variable name ;
- The variable type—which implies how much memory the variable uses;
- The variable address in memory;
- The variable value—this is what fills the bytes at the memory address specified.

> A pointer is used to store the memory address of a variable—it tells you where the variable is. This allows different parts of your code to look at the same variable without having to copy it around.

Here is how we can declare pointers:

```
int *pointVar;
```

Pointer pointVar is declared which can point to a variable of type int. The asterisk is being used to designate a variable as a pointer. Following are the valid pointer declaration:

星号；命名

```
int *pi;        // pointer to an integer
double *pd;     // pointer to a double
float *pf;      // pointer to a float
char *ch;       // pointer to an character
```

The actual data type of all pointers, whether integer, float, character, or any other data type is, the same: a long hexadecimal number that represents a memory address. The only difference between pointers of different data types is the data type of the variable or constant that the pointer points to.

Below is how we use a pointer very frequently:

(1) Define a pointer variable.
(2) Assign the address of a variable to the pointer.
(3) Finally, access the value at the address available in the pointer variable.

This is done by using unary dereference operator (*) that returns the value of the variable located at the address specified by its operand. By preceding the pointer name with the dereference operator, we can get the value which stored in address that the pointer stored. The operator itself can be read as "value pointed to by".

间接引用
前置

（指针变量）指向的地址里的值

Following example makes use of these operations.

```
#include<iostream>
using namespace std;
int main()
{
    int a = 3;         // actual variable declaration.
    int *pa;           // declaring a pointer variable pa
    pa = &a;           // store address of a in pointer variable
    cout << "Value of a variable a: " << a << endl;

    // print the address stored in pa pointer variable
    cout << "Address stored in pa variable: " << pa << endl;
```

```
// access the value at the address available in pointer
cout << "Value of *pa variable: " << *pa << endl;

return 0;
}
```

When the above code is compiled and executed, it produces result something as follows:

```
Value of a variable: 3
Address stored in pa variable: 0x61fe44
Value of *pa variable: 3
```

6.7 Dynamic Array

动态数组

In the programs seen in previous chapters, all memory needed were determined before program execution by defining the variables needed. But there may be cases where the memory needed of a program can only be determined during runtime, for example, when the memory needed depends on user input. On these cases, programs need to dynamically allocate memory by using operator "new". For example:

动态分配内存空间

```
float* pNumber=new float;
```

This makes a new space in memory to hold a single float and pNumber takes the address in memory that the new float is stored at. At first, this float doesn't take any particular value, but we can set it to one of our choosing with the following:

```
*pNumber=10.8;
```

This code changes the value of the float stored at the address in memory given by pNumber.

if a sequence of more than one element is required, the code could be:

```
float* pNumber;
pNumber = new float[5];
```

or:

```
float* pNumber = new float[5];
```

In this case, the system dynamically allocates space for five elements of type float and returns a pointer to the first element of the sequence, which is assigned to pNumber. Therefore, pNumber points to the block of memory with space for five elements of type float (see Figure 6-2).

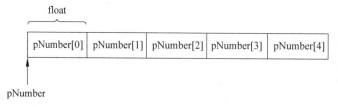

Figure 6-2　Dynamically allocated memory spaces

Here, pNumber is a pointer, and thus, the first element pointed to by pNumber can be accessed either with the expression pNumber[0] or the expression *pNumber (both are equivalent). The second element can be accessed either with pNumber [1] or *(pNumber +1), and so on...

指针

The important difference between a regular array and a dynamically allocated array is that the size of a regular array needs to be a constant expression, and thus its size has to be determined at the moment of designing the program. The dynamic memory allocation performed by a "new" keyword, allows to assign memory during runtime using any variable value as size. So, the code below is allowed:

常规数组
常量表达式
确定了

```
int n = 5;
float* pNumber;
pNumber = new float[n];
```

Incidentally, if you create a pointer to a new area of memory using the new keyword, you must remember to release that memory at some point. Otherwise, it will not be available to other programs until possibly after your PC is rebooted. To free memory used by pNumber, we use the following statement:

释放内存空间

delete pNumber;　　// Releases the memory of a single element allocated using new

delete [] pNumber; // Releases the memory allocated for arrays of elements using new

This allows the float sized region of memory to be made available for reallocation. The example below shows the use of a dynamic array.

```
#include<iostream>
#include<new>
using namespace std;
int main()
{
   int n, i;
   float * pNumber;
   cout << "How many numbers would you like to type? ";
   cin >> n;
   pNumber = new float[n];
   if (pNumber == nullptr)
      cout << "Error: memory could not be allocated";
   else
   {
      for (i=0; i<n; i++)
      {
         cout << "Enter number: ";
         cin >> pNumber[n];
      }
      cout << "You have entered: ";
      for (i=0; i<n; i++)
         cout << pNumber[n] << ", ";
      delete [] pNumber;
   }
   return 0;
}
```

Programming Exercises

1. Write a C++ program that will prompt the user to input ten integer values, then find the largest and smallest elements of the array.

2. Take 10 integer inputs from user and store them in an array. Then ask user to give a number. Now, tell user whether that number is present in array or not.

3. Write a C++ program that will prompt the user to input ten integer numbers,

print the sum, average of all numbers.

4. Write a C++ program to sort 10 integer values (reading from keyboard) in ascending and descending order.

5. Write a piece of code which finds out how many digit characters are in an array.

```
int main()
{
    char arr[5] = {'a', '3', 'n', '5', 'T'};
    int digit_character;
    // continue here
```

6. Write a C++ program to reverse the element of an integer array.

7. Write a menu driven C++ program with following option.   菜单驱动的
   a. Accept elements of an array
   b. Display elements of an array
   c. Sort the array using insertion sort method
   d. Sort the array using bubble sort method

Chapter 7   Functions

Modularization is a very important concept in software design. According to the software engineering, a complex software should be divided into different modules, each module may be developed by different people, even in different part of the world. After each module is finished, they are assembled together. For the beginners in C/C++ programming, the modularization concept is very important. If you need to perform several tasks in a program, you may put all the statements in main(), however, a better way is to write a function for each task, then you call these functions in main( ). The benefits of doing that are:

模块化
软件工程

装配

(1) Make the program more readable;
(2) Reuse some code;
(3) Reduce the chance of making mistakes;
(4) Ease to maintain.

更加易读
重用某些代码

易于维护

7.1   What is a Function

函数

> A function is a group of statements that perform a specific task.

特定任务

A function may take inputs or no input, do some specific computation and produces output. There are two types of function:

- Standard library functions: Predefined in C/C++;
- User-defined function: Created by users.

库函数；预定义

7.2   Library Functions

Library functions are the built-in functions that the compiler has already written for us. You can use library functions by including the corresponding header file.

内建的函数，预先定义好的函数；头文件

```
#include<iostream>
#include<cmath>    //including header file cmath
using namespace std;
int main()
{
   cout << "The square root of 2.0 = ";
   cout << sqrt(2.0) << endl;

   return 0;
}
```

The output is:

```
The square root of 2.0 = 1.41421
```

In example above, sqrt( ) is a library function that used to calculate the square root of a number. In order to use that function, just add

平方根

```
#include<cmath>
```

at the beginning of the file. Here, "cmath" is a header file. The function definition of sqrt( )(body of that function) is written in that file. You can use all functions defined in cmath when you include the file cmath.

For C programming language, the instruction is different:

```
#include<math.h>
```

7.3 User-defined Functions

用户定义的函数

More often, we need to write our own functions, that is called use-defined functions. Functions are useful for encapsulating common operations in a single reusable block, ideally with a name that clearly describes what the function does.

封装

可重用的

7.3.1 Defining a Function

定义函数

A C or C++ function consists of a function header and a function body.

函数头；函数体

The function header, which is the first line of the function, declares the return type, function name, and parameter list (which may be empty) of the function.

Follow the function header is the function body, which is some statements surrounded by a pair of curly braces. The function body defines what the function does.

The general form of a C++ function definition is as follows:

```
return_type function_name(parameter list)    // Function header
{
   // Function body
}
```

Return Type: The data type of the value that the function returns. Some functions perform the desired operations without returning a value. In this case, the return_type should be void.

返回值类型

Function Name: The name of the function. This is the actual name of the function. The function name and the parameter list together constitute the function signature.

函数名
被看作
特征

Parameters: A parameter is like a placeholder. When a function is called, a value is passed to the parameter. This value is referred to as actual parameter. The parameter list refers to the type, order, and number of the parameters of a function. Parameters are optional, that is, a function may contain no parameters.

参数：占位符
实际参数
类型、顺序

Function Body: Braced with a pair of curly brackets, the function body contains a collection of statements that define what the function does.

函数体

Following is the source code for a function called addition(). This function takes two parameters num1 and num2 and returns the sum of the two:

```
// function returning the sum of two numbers
int sum(int num1, int num2)
{
   int sum;            // local variable declaration
   sum = num1 + num2;
   return sum;
}
```

7.3.2 Parameters and Arguments

In common usage, the terms parameter and argument are often used interchangeably. However, for the purposes of further discussion, we will make a distinction between the two.

互换

> A parameter (sometimes called a formal parameter) is a variable declared in the function declaration.
> An argument (sometimes called an actual parameter) is the value that is passed to the function by the caller.

形式参数

实际参数

The example below is another version of addition (), it takes two parameters num1 and num2, so num1 and num2 are the formal parameters.

```
int addition(int num1, int num2)
{
    return (num1 + num2);
}
```

If the function addition () is called in the main(), the code may look like this:

```
int main()
{
  int a = 3, b = 5, c;
  c = addition(a,b);
  cout << "The value of c = " << c << endl;

  return 0;
}
```

The variables a and b are arguments, or actual parameters.

When a function is called, all of the parameters of the function are created as variables, and the value of the arguments are copied into the parameters.

For example above, when addition(int num1, int num2) is called with arguments 3 and 5, parameter num1 is created and assigned the value of 3, and parameter num2 is created and assigned the value of 5.

Even though parameters are not declared inside the function block, function parameters have local scope. This means that they are created when the function is invoked, and are destroyed when the function block terminates.

本地作用范围
激活；销毁；终
止运行

7.3.3 Function Declarations

函数的声明

Usually, a program has more than one functions: some are user-defined functions and one is main() function. As we have already known that no matter the order in which they are defined, a C/C++ program always starts with main( ).

For a C/C++ program, the user-defined functions can write before the main() function or after the main() function. No matter where the user-defined functions are placed, the execution of the program always start with main() function. The main() function is the only function that called automatically by the operating system, and any other functions are only executed when they are called.

If a user-defined function is defined after the main() function, the main() function may not "know" these functions. So in this case, you should tell the main() that there is such a user-defined function. This is called function declaration.

A function declaration tells the compiler about a function name and how to call the function. The actual body of the function can be defined separately. A function declaration has the following parts:

```
return_type function_name(parameter list);
```

For the function addition(), following is the function declaration:

```
int addition(int num1, int num2);
```

In a function declaration, parameter names are not necessary, only the type is required, so following is also valid:

```
int findmax(int, int);
```

Function declaration is also required when you define a function in one source file and you call that function in another file.

7.4 Calling a Function

As we have already known that no matter how many functions there are in a program and the order in which they are defined, a C/C++ program always starts with main().

When a function is called, program control is transferred to this function. It performs defined task and when its return statement is executed or when its function-ending sign—closing brace is reached, it returns program control back to the program which call it.

7.4.1 Passing by Value

> There are three methods of calling a function: pass by value, pass by reference, and pass by address.

When an argument is passed by value, the argument's value is copied into the value of the corresponding function parameter. In this case, changes made to the parameter inside the function have no effect on the argument. For example:

```
#include<iostream>
using namespace std;

void fun(int a, int b)
{
   a *= a;
   b *= a;
   cout << "\ta = " << a << "\tb = " << b << '\n';
}

int main()
{
   int x = 3, y = 5;

   cout << "Initial values of x and y:";
   cout << "\tx = " << x << "\ty = " << y << '\n';

   cout << "Calling the function fun(x,y):";
```

```
    fun(x,y);                    // Calling function fun(x,y)

    cout << "The new values of x and y:";
    cout << "\tx = " << x << "\ty = " << y << '\n';

    return 0;
}
```

The output is:

```
Initial values of x and y are:    x = 3    y = 5
Calling the function fun(x,y):    a = 9    b = 25
The new values of x and y are:    x = 3    y = 5
```

As can be seen, after the calling of the function fun(), the value of arguments x and y are still 3 and 5.

7.4.2 Passing by Address (or Pointers)

传址调用

This method of function call copies the address of an argument into the formal parameter. Inside the function, the address is used to access the actual argument used in the caller. This means that changes made to the parameter affect the argument. Accordingly, the function parameters should be pointer type so as to accept pointers or address. For example:

调用者

```
#include<iostream>
using namespace std;

void swap(int* n1, int* n2) //declaring pointer type parameters
{
    int temp;
    temp = *n1;
    *n1 = *n2;
    *n2 = temp;
}

int main()
{
    int a = 3, b = 5;
    cout << "Before swap() called:";
    cout << "\ta = " << a << ",b = " << b << endl;
```

99

```
    swap(&a, &b);          //address of a and b is passed
    cout << "After swap() called:";
    cout << "\ta = " << a << ",b = " << b << endl;

    return 0;
}
```

In the example above, the address of a and b are passed to the swap() function using swap(&a, &b); Pointers n1 and n2 accept these arguments in the function definition.

When *n1 and *n2 are interchanged inside the swap() function, a and b inside the main() function are also interchanged, because n1 and a are share the same address, and n2 and b share the same address. Notice that swap() is not returning anything; its return type is void. Another example:

```
#include<iostream>
using namespace std;

void add(int* ptr)
{
    (*ptr)++; // adding 1 to *ptr
}

int main()
{
    int* p, a = 2;
    p = &a;
    add(p);
    cout << *p << endl;

    return 0;
}
```

In the program above, p stored the address of a, so *p is the value of a.

Initially the *p is 2. When the pointer p is passed to the add() function, the ptr pointer gets its address.

Inside the function add(), the value stored at ptr is increased by 1 using (*ptr)++. Since ptr and p pointers have the same address, *p inside main() is also increased by 1.

7.4.3 Passing by Reference

A reference variable is a nickname, or alias, for a variable.

To declare a reference variable, we use the unary operator &:

```
int a = 3;        // declares a variable a and assign 3 to a;
int & r = a;      // declares r as a reference to a
```

In this example, r is a reference to a. That means that a and r are referring to the SAME storage location in memory. For example:

```
#include<iostream>
using namespace std;

int main()
{
   int a=3;
   int &r = a;      // r is now a nickname for a

   cout << "a = " << a << " and r = " << r << endl;
   r = 5;           // changing r will change x.
   cout << "a = " << a << " and r = " << r << endl;

   return 0;
}
```

In the program above, a and r have the same value.

Passing by reference means that the memory address of the variable (a pointer to the memory location) is passed to the function. This is much like passing by address, where the address of a variable is passed on.

This method copies the reference of an argument into the formal parameter. Inside the function, the reference is used to access the actual argument used in the call. This means that changes made to the parameter affect the argument.

For passing by reference method, no copy is made, so overhead of copying (time, storage) is saved.

7.5 Recursion

递归（调用）

A function that calls itself is known as a recursive function. And, this technique is known as recursion.

The recursion continues until some condition is met.

To prevent infinite recursion, if...else statement can be used where one branch makes the recursive call and the other doesn't. For example:

```cpp
#include<iostream>
using namespace std;

int factorial(int n)
{
    if (n == 0 || n == 1)
        return 1;
    else
        return n * factorial(n - 1);
}

int main()
{
    int n, factor;
    cout << "Enter a non-negative number: ";
    cin >> n;
    factor = factorial(n);
    cout << "Factorial of " << n << " = " << result;

    return 0;
}
```

The output is:

```
Enter a non-negative number: 5
Factorial of 5 = 120
```

The process of execution of the program above is as below (see Figure 7-1):

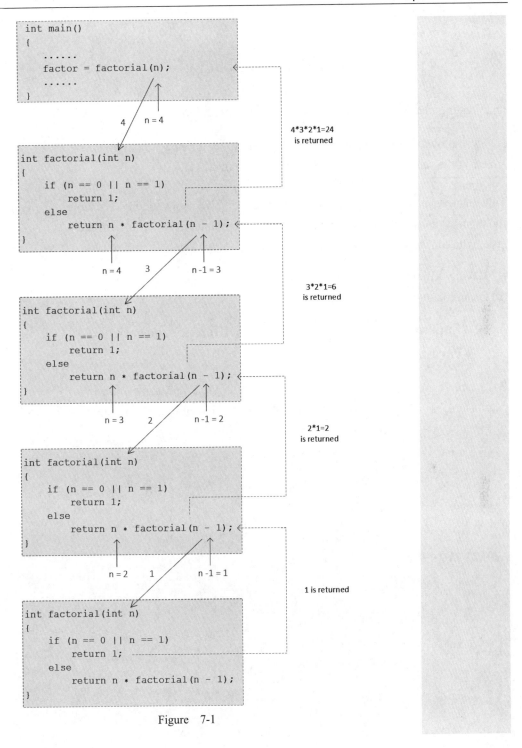

Figure 7-1

7.6 Function Overloading

函数重载

Overloading means the use of the same thing for different purposes.

> C/C++ allows to specify more than one definition for one function name in the same scope, which is called function overloading.

Function overloading is a most important feature of C/C++ in which two or more functions can have the same name but different parameters for the different tasks. The function overloading can be done in two ways:

(1) Having different types of parameters.
(2) Changing number of parameters.

7.6.1 Function Overloading by Having Different types of Parameters

For example, we define two functions with same name and same number of parameters, but the type of the parameter is different. The program is as below.

```
#include<iostream>
using namespace std;
int addition(int a,int b)
{
    return(a+b);
}

double addition(double a,double b)
{
    return(a+b);
}

int main()
{
    cout << addition(2,3) << endl;
    cout << addition(3.2, 5.3) << endl;
    return 0;
```

}

In this program, we have two addition() function, the first one gets two integer parameters, the second one gets two double parameters. The output is :

```
5
8.5
```

7.6.2 Function Overloading by Having Different Number of Parameters

In the code below, two functions addition( ) have the same name and the same type, but they have different number of parameters.

```
#include<iostream>
using namespace std;
int addition(int a,int b)   //This function has two parameters
{
    return(a+b);
}

int addition(int a,int b,int c)  //This function has three parameters
{
    return (a+b+c);
}

int main()
{
    cout << addition(2,3) << endl;
    cout << addition(2,3,4) << endl;
    return 0;
}
```

The figure 7-2 shows the comparison of function overloading with different number of parameters.

Advantages of function overloading in C++:

(1) Performs different operations and hence eliminates the use of different function names for the same kind of operations.

(2) Program becomes easy to understand.

(3) Easy maintainability of the code.

(4) Makes the execution of the program faster and decreases the execution time. 减少执行时间

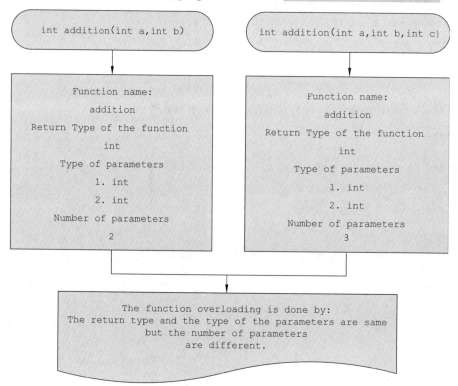

Figure 7-2 The comparison of function overloading with different number of parameters

Programming Exercises

1. Write the output of the following program.

```
#include<iostream>
using namespace std;

void X(int &A, int &B)
{
    A = A + B;
    B = A - B;
    A = A - B;
```

```
}
int main()
{
    int a = 4, b = 18;
    X(a,b);
    cout << a << ", " << b;
    return 0;
}
```

2. Write the output of the following program.

```
#include<iostream>
using namespace std;
int global = 10;
void func(int &x, int y)
{
    x = x - y;
    y = x * 10;
    cout << x << ", " << y << '\n';
}
int main()
{
    int global = 7;
    func(global, global);
    cout << global << ", " << global << '\n';
    func(global, ::global);
    cout << global << ", " << global << '\n';
    return 0;
}
```

3. Write a function that will take three arguments from the caller, then it will return the smallest and largest of those numbers.

4. Write a function that receives two numbers as an argument and display all prime numbers between these two numbers. Call this function from main( ).   素数

5. Write a function to calculate the area of a triangle, it takes three floating point numbers which represent the three sides of the triangle as arguments. Call this function from main( ) and print the results in main( ).

6. Write a program that lets the user perform arithmetic operations on two

numbers. Your program must be menu driven, allowing the user to select the operation (+, −, *, or /) and input the numbers. Furthermore, your program must consist of following functions:

(1) Function showChoice: shows the options to the user and explains how to enter data.

(2) Function add: accepts two number as arguments and returns the sum.

(3) Function subtract: accepts two number as arguments and returns their difference.

(4) Function multiply: accepts two number as arguments and returns the product.

(5) Function divide: accepts two number as arguments and returns the quotient.

Part 3 Object Oriented Programming

Chapter 8 Strings

Without strings, most software would be incapable of doing anything useful. You wouldn't be able to give files names that you could remember, your address book would be filled with numbers that somehow referred to people and everything would be devoid of human readability and therefore largely useless. For this reason, we really need to address the issue of strings, or to give them their full title, character strings.

字符串

没有用的，可读性（人类能够读懂的）

C++ provides following two types of string representations:

- The C-style character string.
- The string class.

C 风格的字符串
String 类

The C-style character string originated from the C language and continues to be supported within C++. This string is actually a one-dimensional array of characters which is terminated by a null character '\0'.

源于
一维数组
空字符

The following declaration and initialization create a string consisting of the word "Shanghai". To hold the null character at the end of the array, the size of the character array containing the string should be one more than the number of characters in the word "Shanghai".

```
char location[9] = {'S','h','a','n','g','h','a','i','\0'};
```

If you follow the rule of array initialization, then you can write the above statement as follows:

```
char location[] = "Shanghai";
```

You may have some previous experience with using character strings in your programs. This was most likely to be of the form:

```
char* name="Shanghai";
```

We know this as a character pointer. This wasn't especially easy to work with for a number of reasons:

- When we passed the variable "name" around, we also had to track the length of the string it represented.
- We had to keep track of the memory used and delete it when we were finished with it.

Luckily, people have seen this kind of thing and have acted to avert future software development slowing to a crawl. As we've noted that C++ was designed to give programmers the speed and flexibility of C but with reusable semantics.

One significant development was the standard template library (STL) which provided third party programmers (like ourselves) with higher level data structures and functions that allowed complex programming tasks to be completed in significantly fewer lines of code than that was previously achievable with procedural languages such as C. In this lesson, the data structure that is of interest to us is string.

字符型指针

避免

可重用的语义

标准模板库
第三方

可以实现的
过程化语言

Here's the simplest example of it in use:

```
1    #include<iostream>
2    #include<string>
3    using namespace std;
4
5    //example 1 demonstrating literal assignment
6    int main(int argc, char** argv)
7    {
8        //assign the literal value to the variable named greeting;
9        string greeting="hello!";
10       cout << greeting << endl;
11       return 0;
12   }
```

Some of the above will seem familiar, but there will be some parts that will be new. Let's step through them one by one.

Before going any further, from now on all of your programs must contain the following statement:

```
#include<string>
```

Then you can create a string variable by adding the statement: string greeting;

Other than that, it's pretty straightforward. The program just prints out the value of the variable "greeting" to the command line then exits. As always, when we declare a variable in C++ we need to choose which type it will be (based on what we are going to do with it – we are choosing string), we need to give it a sensible name (ours is called "greeting" as it represents what we might say when we meet people) and optionally we may want to give it a value. The value can be a literal one as we've done here with "hello!" or it can take the value of another variable of the same type. How do we do this? Read on…

退出

有含义的名字

8.1 String Assignment

字符串赋值

String values can be assigned to a string variable (or string object) in two ways. By assigning a literal value, the code is as below:

```
string greeting1 = "Hello World!";
```

Or by assigning the value of another variable of type string:

```
string greeting1 = "Hello, World!";
string greeting2 = greeting1;
cout << "greeting 1 is: " << greeting1 << endl;
cout << "greeting 2 is now: " << greeting2 << endl;
```

8.2 String Comparison

字符串比较

It's inevitable that you will want to compare two strings to see if they are the same. Using more or less the same syntax as we have been using to test for equivalence with other data types such as int and float, we can check to see if two strings are equal or to see if a string equals a literal value. The code below

不可避免的

show how to compare two strings:

```
string greeting="Hello";
string othergreeting="Hi";
string farewell="Goodbye";

if(greeting == othergreeting)//test if the two strings are the same
    cout << "Both greetings are the same." << endl;
if(greeting == "Hello")
    cout << "Both greetings are the same." << endl;
if(greeting != farewell) //test if the two strings are not equivalent
    cout << "The greeting is not the same as the farewell." << endl;
```

The inequivalence operator ('!=') can be used in a similar manner. The equivalence is sensitive to case ('hEllo' is not the same as 'hello') and does not **ignore spaces** ('hello' is not the same 'he llo') or control characters (tabs, new lines etc.). This will not be the only time we will note during this lesson that, if you have been a C programmer, the previous way of doing this was significantly less convenient.

忽略空格

8.3 String Concatenation

字符串连接

There will also be times when you want to stick strings together or split them apart (more about this). Adding one string on to another is known as concatenation and to do this we use the "+" and "+=" operators. These will be quite similar to the way you've been used to with numeric data types. The code is as follows:

```
string greeting="Hello";
string aspace=" ";
string greetee="World";
string hw;

hw=greeting+aspace+greetee;

cout << "The concatenated greeting is: " << hw << endl;
```

In the same way as we use the '+=' operator as shorthand in arithmetic operations, string concatenations can also be simplified. The code is as below:

```
string greeting="Hello";
string aspace=" ";
string greetee="World";
string hw;
cout << "The greeting is: " << hw << "." << endl;
hw+=greeting;
cout << "The concatenated greeting is: " << hw << "." << endl;
hw+=aspace;
cout << "The concatenated greeting is now: " << hw << "." << endl;
hw+=greetee;
cout << "The concatenated greeting is now: " << hw << "." << endl;
```

For a C programmers who will spend many extra hours (and additional lines of code) just to achieve the same thing.

8.4 String Functions

字符串函数

Strings are different from the basic data types like int, float and char, every string variable created has a number of useful things inside it:

基本数据类型

- the characters in the string;
- a numeric value representing the number of characters;
- functions for chopping up or appending other strings, and much more…

分割；添加

This is possible because string is not one of the basic data types but is what is known as a class. We'll cover classes in more detail at a later time as they are an integral part of C++ and indeed object-oriented programming in general. In short, a class is a specification for a data type that permits variables to be aggregated into a single data structure. Furthermore, this data structure can contain functions that allow the variables inside to be manipulated in some way. Classes will be covered in greater depth later. For now, we just need to know how to use them. The use of string class is as below:

面向对象的
一种数据类型的
详细规范；融入
操作

```
string str("this is a string");   //create a string variable named
//"str", and initialize it with "this is a string"
cout << "The length of the string (" << str << ") is " << str.length()
```

```
<< endl;
```

What this piece of code does is to print out the text that is inside the string and calls the function "length" that returns a count of how many characters are inside the string. The most important thing to remember here is the member selection operator ".". This gives you access to whatever functionality string has to offer. If you are dealing with pointers, this has to be replaced with '->' when you wish to call a function of a string instance. The code is as below:

```
string* str=new string("this is a string");
cout << "The length of the string (" << &str << ") is " << str->length() << endl;
delete str;
```

This does exactly the same thing as the first example.

Again we use a function to obtain some of the strings internal data. This time it's the function "at" which returns the character at a given index.

```
string str="string again!";
int len=str.length();
for(int i=0;i<len;i++)
    cout << "Character at position" << i << "is" << str.at(i) << endl;
```

8.5 String Operations

The use of a class to encapsulate string data and functionality provides us with exceptional convenience. If we want to find a given character or even another string within a string, we can use one of a number of functions in string to do so:

- int find(char findthis)
- int find(char findthis, int startat)
- int rfind(char findthis)
- int rfind(char findthis, int startat)

The above function declarations are simplifications of what you would find if you looked in the string header file, but are functionally identical. There are several others (all overloaded functions) offering various other specializations.

However, more than enough can be achieved with just these four. For example:

```
string str="string again!";
int excpos=str.find('!');
int quespos=str,find('?');
cout << "Found an exclamation mark at position: " << excpos << endl;
cout << "Found a question mark at position: " << quespos << endl;
```

What you should see from running the above example is that the find function returns the position where the first occurrence of the specified character was found. The direction of search is from left to right beginning at the start of the string (position 0). The second time the find function is used to find a question mark ('?'), it returns a value of −1. This value is used as a special indicator for the times when the specified character is not found.

出现；指定的字符

The second implementation of the find function includes an additional argument of type int which tells it where to start looking for a given character. The code is as follows:

```
string str="string! again!";
int excpos1=str.find('!');
int excpos2=str.find('!',excpos1+1);
cout << "Found the first exclamation mark at position: " << excpos1 << endl;
cout << "Found the second one at position: " << excpos2 << endl;
```

When the code is running, we can see the output:

Found the first exclamation mark at position: 6

Found the second one at position: 13

Again, we are searching from left to right. In the second call, we start looking at the place after the one where the last occurrence was found.

出现

The example below shows the use of the function rfind or "reverse find". This works in exactly the same way as find did but reverses through the string beginning at the last character.

```
string str="string! again!";
int excpos1=str.rfind('!');
```

```
int excpos2=str.rfind('!',excpos1-1);

cout << "Found first exclamation mark at position: " << excpos1 << endl;
cout << "Found the second one at position: " << excpos2 << endl;
```

Earlier we looked at ways of joining together strings. It's not unthinkable that you'll want to break them apart as well. Fortunately, the designer of the string thought so too and provided a function called substr which extracts substrings from the string it is called on. The code below shows how to extract substrings.

提取

```
string str="string! again!";
int excpos1=str.find('!');
int excpos2=str.find('!',excpos1+1);

string s1=str.substr(0, excpos1);
string s2=str.substr(excpos1, excpos2-excpos1);

cout << "First sub-string is : " << s1 << endl;
cout << "Second sub-string is : " << s2 << endl;
```

The function substr takes two parameters both of type int: the first is the starting position of the extraction; the second is the number of characters to extract. The function will return the extracted string.

The alternative to the string class is to just use a character array as you would in C. This is painful. Think about it:

- Having to retain an additional int variable to keep track of the string length;
- Having to resize the array when concatenating other strings (and also copy between temporary storage while doing it);
- Iterating the length of each array making a pairwise comparison of each character to test for equivalence.

Chapter Review

1. How to use string from the standard template library?

2. Why string is more powerful and easier than a character array?

3. How to test the length of a string?

4. How to compare two string variables?

5. How to concatenate two or more string variables together into one string?

6. How to access the character at a given position of a string?

7. How to extract a string from inside a string?

Programming Exercises

1. Write a program to find the length of string.

2. Write a program to display a string from backward.

3. Write a program to count number of words in a string.

4. Write a program to concatenate one string contents to another.

5. Write a program to compare two strings to find if they are exact equal or not.

6. Write a program to check is a string is palindrome or not. 回文

7. Write a program to find a substring within a string. If found, display its starting position.

8. Write a program to reverse a string.

9. Write a program to convert the uppercase letters in a string to lowercase letters.

10. Write a program to convert the lowercase letters in a string to uppercase letters.

Chapter 9 Classes and Objects

Data is the driving force behind most programs that you'll write. Actually, the execution of a program is the process of adding or subtracting numbers, parsing strings, counting things, and passing different pieces of data around.

We have already learned some basic data types, such as int, float, double, etc. With these data type, together with arrays and functions, you can write many useful programs.

However, with these basic types of data, we're only capable of representing and working with numbers (int, float, double), text (char) and boolean values (bool). These are going to pose a problem to us: if we want to write more and more complex programs: outside of actual numbers, text and true/false information, most of the things that we interact with in the real world can't be easily just stored in just one string, number or boolean values.

9.1 Data Encapsulation

Supposedly, we need to calculate the area and volume of a cuboid, we need to create three variables: length, breadth and height along with the functions calculateArea() and calculateVolume().

However, in C++, rather than creating separate variables and functions, we can wrap these related data and functions in a single unit by creating a class.

Another instance, in order to represent an air ticket, you need many values to store it, such as:

- flight number
- airline
- origin (where it takes off)
- destination (where it lands)

- frequent flyer points
- passenger name
- seat allocation

All of these items of data may require different types with which to represent them, so we can't just make an array and store the values in that.

It is extremely important for us to wrap these related data and functions in a single unit by creating a class, because all these data can work in an organized and structured way, as they all belong to one thing.

C++ provide a very useful tool to solve these problems, that is: object-oriented programming (OOP). OOP allows variables in your programs to interact in a similar way to that of entities in the real world.

With object-oriented programming technique, you could create your own data types by using any fundamental data types (like int, char, float, etc). By this means, you can theoretically model any real-world object or entity and use it just like it would a number or a string. This kind of user defined data type is called class.

A class defines a real-world object in two aspects: its attributes and its behaviour.

For example, a triangle has attributes such as three sides, each side may have different length. For a triangle, we may need to know its perimeter and area, so the behaviour should be calculating perimeter and area.

The class mechanism in C++ provides us a method to create a virtual object by representing its attributes using variables, and representing its behaviours using functions. The variable in a class is called member variable, and a function in a class is called member function.

Class { Member variables – describing the attributes
        Member functions – describing the behaviours

> A class in C++ is a user-defined data type declared with keyword class that has member variables and member functions as its members whose access is governed by three access specifiers private, protected and public.

用户定义的

私有类型
保护类型；公有类型

By default, the member variables are not accessible outside the class; they can be accessed only through member functions of the class. The public members form an interface to the class and are accessible outside the class.

A class combines data and functions in a single unit, this is called encapsulation. It is one of three important features of object-oriented programming.

封装

Why encapsulation? Consider a real-life example, a electric appliance, such as a TV set, has an outer case. It protects the people from electric shocks, it also protects the components or electronic circuits of the TV set from being damaged. The user can operate the TV set by a remote controller.

Encapsulation is also known as data hiding. It makes the variables in a class accessible only by its member functions. Encapsulation improved the security of the data and simplified the programming.

数据隐藏
安全性

9.2 Declaring a Class

类的声明

The syntax for a class declaration is as follows:

```
class class_name
{
private:                      //access modifier
   member variables 1;        //member variables
   member variables 2;
   ...
public:                       //access modifier
   member function 1;         //member functions
   member function 2;
   ...
};
```

Where class_name is a valid identifier for the class. The body of the declaration

有效的

contains member variables and member functions, and access modifiers.

> Access to the class members from outside the class is controlled by "access modifiers". An access modifier can be any of the keywords: public, private or protected.

Public members can be accessed from any place where the class is visible.

Private members can be only accessed by other members of the same class, that is to say, it can not be accessed outside the class.

Protected members can be accessed by other members of the same class, and by members of its subclass.

Programming Tips:

1. Make all member variables private. In fact, by default, all the class members are private, if no access modifier is given.

2. Some member functions may be private. But most member functions are public so as to provide an interface to access or change the private member variables from outside the class.

3. Every class declaration must end with a semicolon after the closing brace.

9.3 Defining a Member Function

A member function is the function of the class and works on the members of the class. The definition of a member function can be inside or outside the definition of class.

If a member function is defined inside the class definition, it can be defined directly. But if it is defined outside the class, then a scope resolution operator (::) should be used along with the class name.

9.3.1 Getter and Setter

As the private member variables can not be accessed from outside the class, some special public member functions must be added to a class in order to assign and retrieve data. 取得，提取

The member functions which used to assign data to the member variables are called setter, and the functions which used to retrieve data from the member variables are called getter.

9.3.2 Implementing Member Functions   成员函数的实现

The member functions that declared in the class need to be provided its definition.

The declaration of a member function is similar to that of a normal function. Each has a function header and a function body.

Figure 9-1 is the structure of a class.

```
class triangle
{
private:
    float side_a;
    float side_b;
    float side_c;
public:
    void set_side_a(float a);
    void set_side_b(float b);
    void set_side_c(float c);
    float get_side_a();
    float get_side_b();
    float get_side_c();
    bool check();
    float calculating_area();
};
```

- programmer-specified class name
- class members can be accessed through the public member functions(data hiding)
- constitute the public interface for the class
- private member variables
- public member functions
- end with a semicolon for the class declaration

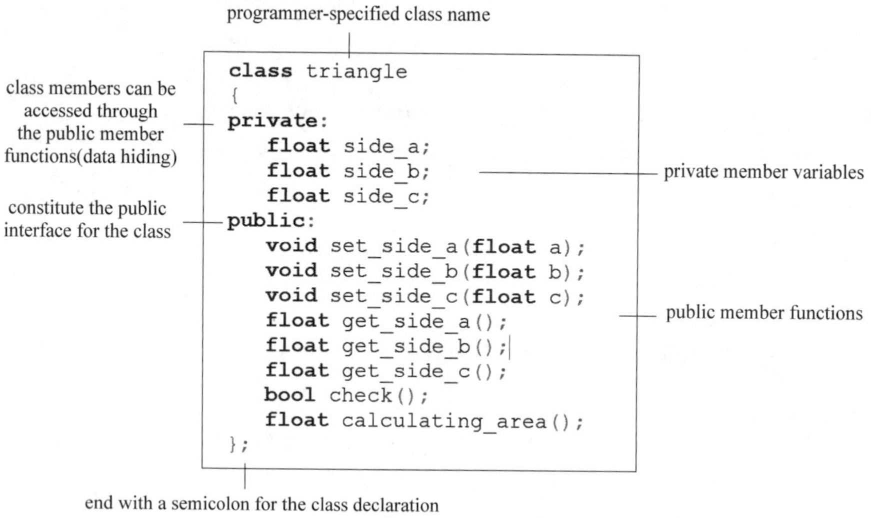

Figure 9-1   The structure of a class

There are two ways to define the member functions: inside the class definition and outside the class definition.

For short member function definitions, you can write the definition inside the class declaration. The code below shows how to define member functions inside a class declaration.

```
class triangle
{
private:
    float side_a;
    float side_b;
    float side_c;
public:
    void set_side_a(float a)
    {
        side_a = a;
    }
    void set_side_b(float b)
    {
        side_b = b;
    }
    void set_side_c(float c)
    {
        side_c = c;
    }
    float get_side_a()
    {
        return side_a;
    }
}
```

In most cases, the member function definitions are written outside the class declaration. If a member function is defined outside the class, a scope-resolution operator (::) need to be placed before the function name to identify the class to which the function belongs. The syntax is as below:

```
ReturnedType className::FunctionName(ParameterList)
{
    FunctionImpeletingStatements;
}
```

For example:

```
void triangle::set_a(float a1)
{
    a = a1;
}
```

9.4 Creating an Object 创建对象

A class is a type that you defined, as opposed to the basic types, such as int and char, that are already defined by the compiler. What we use in a program is the variable of that data type, not the data type itself. Look at the code below:

```
int num;
```

This line of code creates a variable named num of type int. We use num to store integer values.

In the same way, we can use the newly-defined class "triangle" to create variables.

```
triangle triangle1;
```

This line of code creates a variable named triangle1 of the triangle data type. The variables of the class are called objects. An object is an instance of the class. When a class is declared, no memory is allocated. But when an object is created, memory is allocated.

Accessing Member Variables 访问成员变量

A value for an object is a set of values of the member variables. For example, a value for the type triangle consists of three numbers of type float. When accessing a member variable or a member function, a dot (".") operator is used 点运算符 to specify a member. For example, if you want to assign a value of 3.0 to side_a ,then you will have to write:

```
triangle_1.side_a = 3.0;
```

The member functions are also accessed in the same way.

9.5 Constructors for Initialization 构造函数

When an object is declared, we need to assign proper values for some or all the member variables of that object, this is called initialization.

There are two ways to initialize an object:

1. Using constructors.
2. Calling the setter functions.

> A constructor is a special member function that is automatically called when an object of that class is created.

A constructor has some special characteristics:

- A constructor is a public member function of a class that has the same name as the class.
- A constructor definition has no return type, not even void.
- Constructors are used to initialize objects.

If you do not specify a constructor, C++ compiler will generate a **default constructor** for you (expects no parameters and has an empty body).

缺省构造函数（无参）

Default Constructor: Default constructor is the constructor which doesn't take any parameter and doing nothing.

Parameterized Constructors: It is possible to pass arguments to constructors. Typically, these arguments help initialize an object when it is created. To create a parameterized constructor, simply add parameters to it. When you define the constructor's body, use the parameters to initialize the object.

有参构造函数

A constructor can be defined in the same way that you define any other member function.

Here is a class with a constructor:

```cpp
class Triangle
{
private:
    float side_a;
    float side_b;
    float side_c;
public:
    Triangle(float a,float b,float c);//parameterized constructor declaration
    void set_side_a(float a);
    void set_side_b(float b);
    void set_side_c(float c);
    float get_side_a();
    float get_side_b();
    float get_side_c();
    bool check();
    float calculating_area();
};
```

9.6 Destructor

A destructor is a special member function that is called when the lifetime of an object ends. The purpose of the destructor is to free the resources that the object may have acquired during its lifetime.

- Destructor has same name as the class preceded by a tilde (~);
- Destructor doesn't take any argument and don't return anything.

The code below show that a class declaration with a constructor and a destructor.

```cpp
#include<string.h>
#include<iostream>
using namespace std;

class String
{
private:
    char *text;
    int  sizeOfText;
public:
    String(char *ch);        // Declare constructor.
    ~String();               // Declare destructor.
    void printtext();        // member function to print text.
};

String::String(char *ch)    // Define the constructor.
{
    sizeOfText = strlen(ch) + 1;
    // Dynamically allocate the correct amount of memory.
    text = new char[sizeOfText];
    strcpy(text,ch);         // Copy the initialization string.
}

String::~String()           // Define the destructor.
{
    delete[] text;           // Deallocate the memory.
}
```

```cpp
void String::printtext()
{
    cout << text << endl;
}

int main()
{
    String str("Hello");
    str.printtext();
    return 0;
}
```

In the preceding example, the destructor String::~String uses the delete operator to deallocate the space dynamically allocated for text storage.

When do we need to write a user-defined destructor?

If we do not write our own destructor in class, compiler creates a default destructor for us. The default destructor works fine unless we have dynamically allocated memory or pointer in class. When a class contains a pointer to memory allocated in class, we should write a destructor to release memory before the class instance is destroyed. This must be done to avoid memory leakage. 内存泄漏

9.7 Air ticket example for classes and objects

Think about an air ticket – you could not represent one using a single variable as there are too many values to store such as:

- flight number
- airline
- origin (where it takes off)
- destination (where it lands)
- passenger name
- seat allocation

The declaration of an air ticket should be like this:

```cpp
class air_ticket
```

127

```
{
protected:
    string m_passenger;
    int m_flightNumber;
    string m_date;
    string m_origin;
    string m_destination;
    string m_carrier;
public:
    air_ticket();
    air_ticket(string,string);
    ~air_ticket();

    void set_passenger(string name);
    string get_passenger();

    void set_origin(string);
    void set_destination(string);

    string get_origin();
    string get_destination();

    string get_date_of_travel();
    string get_carrier();

    int get_flight_number();
    void set_flight_number(int);

    virtual bool lounge_access();
    virtual int get_seat_pitch();

    string print_ticket();
};
```

The definitions of the member functions is as below:

```
air_ticket::air_ticket()
{
    m_date = "23rd May 2019";
    m_carrier = "KLM";
}
air_ticket::air_ticket(string date,string airline)
```

```
{
    this->m_date=date;
    this->m_carrier=airline;
}
air_ticket::~air_ticket()
{
    //nothing to clean up
}
void air_ticket::set_origin(string org)
{
    this->m_origin=org;
}
void air_ticket::set_destination(string dest)
{
    this->m_destination=dest;
}
void air_ticket::set_passenger(string name)
{
    m_passenger = name;
}
string air_ticket::get_passenger()
{
    return m_passenger;
}
string air_ticket::get_origin()
{
    return this->m_origin;
}
string air_ticket::get_destination()
{
    return this->m_destination;
}
string air_ticket::get_carrier()
{
    return this->m_carrier;
}
void air_ticket::set_flight_number(int num)
{
    this->m_flightNumber=num;
}
int air_ticket::get_flight_number()
```

```
{
    return this->m_flightNumber;
}
```

The definitions of the member functions (continue):

```
bool air_ticket::lounge_access()
{
    return false;
}
int air_ticket::get_seat_pitch()
{
    return 98;
}
string air_ticket::print_ticket()
{
    ostringstream oss;

    oss << endl;
    oss << "==============================" << endl;
    oss << "Name: " << m_passenger << endl;
    oss << "Flight: " << m_carrier << " " << this->m_flightNumber;
    oss << endl;
    oss << endl;
    oss << "From: " << m_origin << endl;
    oss << "To: " <<  m_destination << endl;
    oss << endl;

    oss << "Lounge: ";

    if(lounge_access()==true)
        oss << "Y";
    else
        oss << "N";
    oss << " | Seat Pitch: " << get_seat_pitch() << " cm" << endl;

    oss << "Date: " << m_date << endl;
    oss << "==============================" << endl;
    oss << ends;
    return oss.str();
}
```

For the code above, we use string stream to hold all the printing information. With string stream, we can easily save the printing information to a file.

字符串流

The main() function will create an air ticket object, input some information, and then print out the air ticket. Please pay attention to the member function of print_ticket(), it is type of string, that means you can easily process the ticket information, append it to the other strings, print it out, or save it to a file.

The main() function should be look like this:

```cpp
#include<iostream>
#include<string>
#include<sstream>
using namespace std;
int main()
{
    air_ticket at1;            //create an object of air-ticket
    at1.set_passenger("CZM");
    at1.set_origin("Shangahi");
    at1.set_destination("Beijing");
    at1.set_flight_number(9008);
    cout << at1.print_ticket() << endl;

    string details;
    details=bat.print_ticket();
    cout << details << endl;

    return 0;
}
```

9.8 Friend Function of a Class

类的友元函数

Data hiding is a fundamental concept of object-oriented programming. It restricts the access of private members from outside of the class.

限制访问

But in some real-time applications, in some circumstances, it is more convenient to grant member-level access to functions that are not members of a class or to all members in a separate class.

允许成员级的访问

A friend function, that is a "friend" of a given class, is a function that is defined outside that class, but it has the right to access all private and protected members of the class. Even though the prototypes for friend functions appear in the class definition, friend functions are not member functions of the class.

To declare a function as a friend of a class, precede the function prototype in the class definition with keyword friend, the code is as below:

```
class className
{
    ...
    friend returnType functionName(parameter list);
    ...
}
```

For example, if we have a class of Point, we can calculate the distance of two points, the code can be as below:

```
#include<iostream>
#include<cmath>
using namespace std;

class Point
{
private:
    float xPos; //x coordinate
    float yPos; //y coordinate
public:
    void setPos(float x,float y); //setter function
    float get_x();                //getter function
    float get_y();                //getter function
};
void Point::setPos(float x,float y)
{
    xPos = x;
    yPos = y;
}
float Point::get_x()
{
    return xPos;
}
```

```
float Point::get_y()
{
    return yPos;
}
float distanceOfPoints(Point p1,Point p2)//function which used to
//calculate the distance of two points
{
    return sqrt((p1.get_x()-p2.get_x())*(p1.get_x()-p2.get_x())+
            (p1.get_y()-p2.get_y())*(p1.get_y()-p2.get_y()));
}
int main()
{
    Point p1,p2;
    p1.setPos(0,0);
    p2.setPos(1,1);

    cout << "The distance is: " << distanceOfPoints(p1,p2) << endl;

    return 0;
}
```

For the code above, a function named distanceOfPoints() is used to compute the distance of two points. As it is not the member function of the class Point, it can not access the private members of the class, it has to call the public getter member functions to get the coordinates of the two points.

If the function distanceOfPoints() is declared as a friend function of the class Point, then it would be able to access the private members of the class Point.

```
#include<iostream>
#include<cmath>
using namespace std;

class Point
{
private:
    float xPos;
    float yPos;
public:
    void setPos(float x,float y);
    float get_x();
```

```cpp
        float get_y();
        friend float distanceOfPoints(Point p1,Point p2);//declared as
        //friend function of the class.
};
void Point::setPos(float x,float y)
{
    xPos = x;
    yPos = y;
}
float Point::get_x()
{
    return xPos;
}
float Point::get_y()
{
    return yPos;
}
float distanceOfPoints(Point p1,Point p2)//this function can
//access private members of the class.
{
        Return sqrt((p1.x-p2.x)*(p1.x-p2.x)+(p1.y-p2.y)*
                (p1.y-p2.y));
}
int main()
{
    Point p1,p2;
    p1.setPos(0,0);
    p2.setPos(1,1);

      cout << "The distance is: " << distanceOfPoints(p1,p2) << endl;

      return 0;
}
```

Please pay attention to the definition of the function distanceOfPoints(), private member variables can be accessed directly.

9.9 Operator Overloading

运算符重载

Operator overloading is a specific case of polymorphism. It is one of the nice features of C++. With operator overloading you can give special meanings to

多态

operators when they are used with user-defined classes.

For example, to overload the + operator for your class, you would provide a member function named operator+ on your class.

The following set of operators can be overloaded for user-defined classes:

= (assignment operator)

+ − * (arithmetic operators)

+= −= *= (compound assignment operators)

== != (comparison operators)

For example, for two point objects p1 and p2, they can not add directly by using operator "+", as the C++ do not know how to add two points. By means of operator overloading technique, we can overload the "+" operator in the Point class to support addition of two Point objects. In other words, we can write p3 = p1+p2, which is similar to the usual arithmetic operation. The declaration of the class is as below:

```
class Point
{
private:
    float xPos;
    float yPos;
public:
    void setPos(float x, float y);
    float get_x();
    float get_y();
    void display(); // print out the coordinate of a point.
    // Overload '+' operator as member function of the class.
    Point operator+(const Point & p);
};
```

A new member function is added in the class, the definition of the function is as below:

```
// Member function overloading '+' operator
Point Point::operator+(const Point &p)
```

```
{
    return Point(xPos + p.xPos, yPos + p.yPos);
}
```

To display the coordinate of a point, a new member function display() is added as follows:

```
void Point::display()
{
    cout << "The coordinate is:"
         << "(" << x << ","
         << "y" << ")." << endl;
}
```

In the main() function, we create two point objects, and initialize them ,then we create an object named "p3". We can use "+" to add p1 and p2, and then assign the sum of these two objects to p3. The code is as follows:

```
#include<iostream>
using namespace std;

int main()
{
    Point p1.sePos(1, 2), p2.setPos(3, 4);

    // Use overloaded operator +
    Point p3 = p1 + p2;
    p1.display();    // (1,2)
    p2.display();    // (3,4)
    p3.display();    // (4,6)

    return 0;
}
```

Another example for ++ Operator (Unary Operator) Overloading is as below:

```
// Overload ++ when used as prefix
#include<iostream>
using namespace std;

class charAddition
{
```

```
    private:
        char ch;
    public:
        charAddition(): ch('A') {} // Constructor to initialize ch to A
        void operator ++ () // Overload ++ when used as prefix
        {
            ch = ch + 1;
        }
        void display()
        {
            cout << "ch: " << ch << endl;
        }
};
int main()
{
    charAddition ch1;
    ++ch1;         // Call the "void operator ++ ()" function
    ch1.display();
    return 0;
}
```

The output is:

```
Ch: B
```

Here, when we use ++ch1, the void operator ++ ( ) is called. This increases the ASCII value for the object ch1 by 1, so the output of ch1 should be B.

Chapter Review

1. What is a class?

2. What is an object?

3. What is the relationship between a class and objects that created by the class?

4. What is a constructor?

5. How to overload constructors?

Programming Exercises

1. Write a class with two public member variables that represent the length and width of a the length and width of a rectangle, and one public member function which will return the area of the rectangle.

2. For the exercise 1, change the access modifier of the length and breadth to private, how to modify the program?

3. For the exercise 2, add one default constructor and a parameterized constructor.

4. Write the definition for a class called complex that has floating point data members for storing real and imaginary parts. The class has the following member functions:
   void set (float, float) to set the specified value in object
   void disp() to display complex number object
   complex sum(complex) to sum two complex numbers & return complex number
   (1) Write the definitions for each of the above member functions.
   (2) Write main function to create three complex number objects. Set the value in two objects and call sum() to calculate sum and assign it in third object. Display all complex numbers.

5. Write the definition for a class called time that has hours ,minutes and seconds as integer. The class has the following member functions:
   void settime(int, int) to set the specified value in object
   void showtime() to display time object
   time sum(time) to sum two time objects & return time
   (1) Write the definitions for each of the above member functions.
   (2) Write main function to create three time objects. Set the value in two objects and call sum() to calculate sum and assign it in third object. Display all time objects.

6. Write a program to calculate the area of a triangle.

7. Write a three-dimensional point class named Point_3D, then write a function with two parameters of type of Point_3D to calculate the distance of two points.

8. For the exercise 7, calculate the distance of two points, using a friend function.

9. Write a business_air_ticket class, it has private members below:

```
string m_date;              //what date we fly on
int m_flightNumber;         //what the flight number is
string m_carrier;           //who is taking us
string m_origin;            //where we take off from
string m_destination;       //where we land
string m_passenger;         //the name of the passenger
int m_frequentFlyerPoints;  //flyer points from the carrier
```

Chapter 10　Inheritance

Nobody likes doing more work than they have to. Before object-oriented programming was invented, reusing code meant copying and pasting from one source file into another, then changing it to suit the application. This creates more code (which we have always said creates more trouble) and also doesn't take modifications to the original code into account: if you find a bug in one piece of code, you need to find all the places you pasted that code and fix it again, and again, and again. You'll be glad to hear that this is no longer necessary; OOP allows us to write code once and use it over and over again. If we don't like what the code does, we can change part of it and keep the part we do like.

Inheritance allows the users to create a new class from an existing class. It is one of the most important features of object-oriented programming in C++. The class being inherited from is called the base class, or superclass; The class doing the inheriting is called the derived class, or subclass.

继承

基类；父类

派生类；子类

10.1　Implementing Inheritance

实现继承

For creating a sub-class which is inherited from the base class, we have to follow the below syntax:

```
class subclass_name : access_mode base_class_name
{
  //body of subclass
};
```

Here, subclass_name is the name of the sub class, and base_class_name is the name of the base class.

Access_mode is the mode in which you want to inherit this sub class: public, protected or private. Public inheritance is commonly used. When deriving a class

访问属性

from a public base class, public members of the base class become public members of the derived class and protected members of the base class become protected members of the derived class. Base class's private members are never accessible directly from a derived class, but can be accessed through calls to the public and protected members of the base class.

A derived class inherits all the member variables and functions from the base class with the following exceptions:

1. Constructors, destructor and copy constructors of the base class.

2. Overloaded operators of the base class.

3. The friend functions of the base class.

```
#include<iostream>
using namespace std;

class Animal      // base class
{
public:
    void run()
    {
        cout << "I can run!" << endl;
    }
    void sleep()
    {
        cout << "I can sleep!" << endl;
    }
};

class Dog : public Animal    // derived class
{
public:
    void bark()
    {
        cout << "I can bark! Woof woof!!" << endl;
    }
};

int main() {
```

```
    Dog dog1;          // Create an object of the Dog class
    dog1.run();        // Calling members of the base class
    dog1.sleep();
    dog1.bark();       // Calling member of the derived class

    return 0;
}
```

The output is:

```
I can run!
I can sleep!
I can bark! Woof woof!!
```

In this example, Dog is a subclass to Animal. Dog inherits from animal, acquires behaviors (member functions) from that of Animal. Additionally, Dog defines one member function of its own: bark(). This property is specific to a Dog, not to any Animal.

10.2 Types of Inheritance

继承的类型

In C++, we have 5 different types of inheritance. They are:

(1) single inheritance.

单继承

(2) multiple inheritance.

多继承

(3) hierarchical inheritance.

分支继承

(4) multilevel inheritance.

多层继承

(5) hybrid inheritance (also known as virtual inheritance).

混合继承

1. Single Inheritance

In this type of inheritance, the derived class inherits from only one base class. It is the most simple form of inheritance. See Figure 10-1.

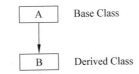

Figure 10-1   Single inheritance

2. Multiple Inheritance

Where one class can have more than one superclass and inherit features from all parent classes. See Figure 10-2.

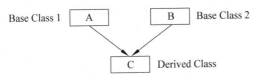

Figure 10-2   Multiple inheritance

3. Hierarchical Inheritance

In this type of inheritance, multiple derived classes inherit from a single base class. See Figure 10-3.

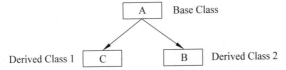

Figure 10-3   Hierarchical inheritance

4. Multilevel Inheritance

In this type of inheritance, the derived class inherits from a class, which in turn inherits from some other classes. See Figure 10-4.

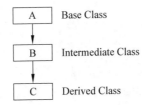

Figure 10-4   Multilevel inheritance

5. Hybrid Inheritance

Hybrid inheritance is combination of hierarchical and multilevel inheritance. See Figure 10-5.

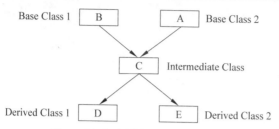

Figure 10-5 Multilevel inheritance

10.3 Access Modes in C++ Inheritance

Depending on access modifier used while inheritance, the availability of class members of super class in the sub class changes. It can either be public, protected or private.

1. Public Inheritance

公有继承

When public inheriting from a base class, public members of the base class become public members in the derived class, protected members of the base class become protected members in the derived class. A base class's private members are never accessible directly from a derived class, but can be accessed through calls to the public and protected members of the base class.

2. Protected Inheritance

保护继承

When protected inheriting from a base class, public and protected members of the base class become protected members of the derived class.

3. Private Inheritance

私有继承

When protected inheriting from a base class, public and protected members of the base class become private members of the derived class.

The table 10-1 summarizes the above three modes and shows the access specifier of the members of base class in the sub class when derived in public, protected and private modes.

Table 10-1 Access properties for different type of inheritance

Base class member access specifier	Type of inheritance		
	public	protected	private
public	public	protected	private
protected	protected	protected	private
private	Not accessible	Not accessible	Not accessible

10.4 Example for Inheritance

We have learned how to write a two-dimensional point class in the previous chapter. In this chapter, we try to write a two-dimensional class and then derive a three dimensional point class. The declaration of a two-dimensional point class looks like this:

```
class Point_2D
{
private:
    float xPos;
    float yPos;
public:
    void setPos(float x,float y);
    float get_x();
    float get_y();
    void dispaly();
};
```

For a three-dimensional point class, it can public inherit from the two-dimensional point class.

```
class Point_3D : public Point_2D
{
private :
    float zPos;
public :
    void setPos(float z);
    float get_z();
```

145

```
    friend float distance(Point_3D p1, Point_3D p2);
};
```

In order to make the private member variables can be accessed by the derived class, the access modifier of the member variables of base class should be protected.

```
class Point_2D
{
protected:            // change "private" to "protected".
    float xPos;
    float yPos;
public:
    void setPos(float x,float y);
    float get_x();
    float get_y();
    void dispaly();
};
```

10.5 Air Ticket Example for Inheritance

Real-world entities, like an air ticket, can be naturally described by both data and functionality. If you remember, the economy air ticket class declaration looked like this:

```
class air_ticket
{
private:
    string m_name;              //who take the airplane
    string m_date;              //what date we fly on
    int m_flightNumber;         //what the flight number is
    string m_carrier;           //who is taking us
    string m_origin;            //where we take off from
    string m_destination;       //where we land
public:
    air_ticket(string,string);
    ~air_ticket();

    void set_origin(string);
    void set_destination(string);
```

```
    string get_origin();
    string get_destination();

    string get_date_of_travel();
    string get_carrier();

    int get_flight_number();
    void set_flight_number(int);

    bool lounge_access();      //do we get lounge access?
    int get_seat_pitch();      //how much leg room do we get?

    string print_ticket();
};
```

Air tickets come with various features depending on how much money you pay for them. The most obvious specialization is business class. This type of ticket has all the functionality of an air ticket in that it gets you from an origin to a destination, but it also lets the passenger collect frequent flyer points (or 'air miles' as they are sometimes called). In order to make this specialization of the air_ticket class, we can create a new class that inherits all of the air_ticket functionality and attributes as follows:

专门化

```
class business_air_ticket : public air_ticket
{
private:
    int m_frequentFlyerPoints;
public:
    business_air_ticket(string,string);
    ~business_air_ticket();
};
```

We now know that inheritance allows us to tailor a class to our requirements by adding member variables and functions to get additional functionality—we call this specialization. But specialization goes further than just adding more data and functions to existing classes. It also allows functionality of the methods of the superclass to be replaced in the subclass using a mechanism known as overriding. All this does is provides the subclass with a new body for the

裁减，修改

覆盖

147

function with a given argument list, name and return type (you have to provide exactly the same declaration otherwise it will not work). For example, in the air_ticket class, we have the following two functions:

```
bool lounge_access();
int get_seat_pitch();
```

If we provide business_air_ticket with the same function declarations like this:

```
class business_air_ticket : public air_ticket
{
private:
    int m_frequentFlyerPoints;
public:
    business_air_ticket(string,string);
    virtual ~business_air_ticket();

    int get_frequent_flyer_points();
    void set_date(string);

    bool lounge_access();
    int get_seat_pitch();
};
```

Then we can implement the above functionality for the class definition:

```
business_air_ticket::business_air_ticket(string passenger,
    string date,string carrier)
{
    m_passenger = passenger;
    m_date = date;
    m_carrier = carrier;
}

business_air_ticket::~business_air_ticket()
{
}

bool business_air_ticket::lounge_access()
{
    return true;                    // becomes true,
}
```

```
int business_air_ticket::get_seat_pitch()
{
    return 155;              // becomes 155.
}
```

Here is a code for main function:

```
int main()
{
    business_air_ticket bat("CZM","20th September 2019","KLM");
    bat.set_origin("Amsterdam Schiphol");
    bat.set_destination("Helsinki Vantaa");
    bat.set_flight_number(1171);

    bat.print_ticket();

    return 0;
}
```

When we execute the program, however, it doesn't quite print out what we expect:

```
==============================
Passenger: CZM
Flight: KLM 1171

From: Amsterdam Schiphol
To: Helsinki Vantaa

Lounge: N | Seat Pitch: 98 cm
Date: 20th September 2019
==============================
```

It's just the same as it was for the air_ticket class — the seat pitch and lounge access values are the ones for an air_ticket instance. Why?

For the business_air_ticket, the print_out() function is derived from the base class air_ticket. The base class air_ticket also have the same functions called lounge_access() and get_seat_pitch(). When the print_out() function of the derived class is called, the print_out() function will call the member functions in the superclass, not the member functions of the derived class.

In order to solve this problem, virtual function is introduced.

虚函数

A virtual function is a member function which is declared within a base class and is re-defined (overriden) by a derived class. When a derived class object calls this function, the derived class's version of the function is executed.

覆盖(重写)

In order to make the main() function work properly, we need to modify the declaration of the base class as follows:

```
class air_ticket
{
private:
    ...
public:
    air_ticket(string,string);
    ~air_ticket();

    void set_origin(string);
    void set_destination(string);

    string get_origin();
    string get_destination();

    string get_date_of_travel();
    string get_carrier();

    int get_flight_number();
    void set_flight_number(int);

    virtual bool lounge_access();   //add virtual keyword before
                                    //the function declaration
    virtual int get_seat_pitch();

    string print_ticket();
};
```

Keyword "virtual" is added before the lounge_access() and get_seat_pitch(). When the print_out() function of the derived class (business_air_ticket) is called, the definition which defined in the derived class will be called. When the program is executed, the output would be:

```
==============================
Passenger: CZM
Flight: KLM 1171

From: Amsterdam Schiphol
To: Helsinki Vantaa

Lounge: Y | Seat Pitch: 155 cm
Date: 20th September 2019
==============================
```

10.6 Polymorphism in C++

Polymorphism is part of the features of object-oriented programming. In C++, polymorphism causes a member function to behave differently based on the object that calls it. Polymorphism is a Greek word that means to have many forms. 　希腊语单词

For Example, the "+" operator is used to perform addition of two numbers, see code below:

```
int a=3, b=4, c;
c=a+b;
cout << "c=" << c << endl;
```

But, when we use "+" operator with the string objects, it performs string concatenation.

For Example:

```
string str1 = "Hello";
string str2 = "world!";
string str3 = str1 + str2;    // str3 = "Hello world!"
```

Types of Polymorphism (Figure 10-6) 　多态的类型

C++ supports two types of polymorphism:

- Compile-time polymorphism; 　编译时多态

- Runtime polymorphism.

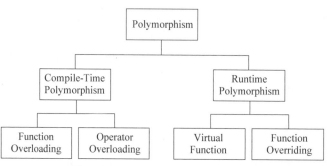

Figure 10-6 Types of polymorphism

1. Compile-time Polymorphism

The overloaded functions are invoked by matching the number and type of arguments. The information is present during compile-time. This means the C++ compiler will select the right function at compile time.

Compile-time polymorphism is achieved through function overloading and operator overloading.

Function overloading occurs when we have many functions with the same names but different arguments. The arguments may differ in terms of number or type.

2. Runtime Polymorphism

This happens when an object's function is called during runtime rather than during compile-time. Runtime polymorphism is achieved through function overriding. The function to be invoked is established during runtime.

In case of function overriding, we have two definitions for the same function name, one is for parent class and the other one is for the child class. The call to the function is determined at runtime, that's the reason why it is called runtime polymorphism. For example:

```
// Example of Runtime Polymorphism
#include<iostream>
using namespace std;
class A
{
public:
```

```cpp
        void display()
        {
            cout << "Super Class Function is called." << endl;
        }
};

class B: public A
{
public:
    void display()
    {
        cout << "Sub Class Function is called." << endl;
    }
};

int main()
{
    A obj1;         //Parent class object
    B obj2;         //Child class object
    obj1.display();
    obj2.display();
    return 0;
}
```

Class A has a function called display, and the derived class B also has a function called display. For the objects of derived class, when member function display is called, the definition inside the derived class will be executed.

Chapter Review

1. What is a derived class?

2. How many types of inheritance?

3. Answer the questions (1) to (4) based on the following:

```cpp
class Publisher
{
    char pub[12];
    double turnover;
```

```cpp
    protected:
        void register();
    public:
        Publisher();
        void enter();
        void display();
};

class Branch
{
    char city[20];
    protected:
        float employees;
    public:
        Branch();
        void haveit();
        void giveit();
};

class Author : private Branch, public Publisher
{
    int acode;
    char aname[20];
    float amount;
    public:
        Author();
        void start();
        void show();
};
```

(1) Write the names of data members, which are accessible from objects belonging to class Author.

(2) Write the names of all the member functions which are accessible from objects belonging to class Branch.

(3) Write the names of all the members which are accessible from member functions of class Author.

(4) How many bytes will be required by an object belonging to class Author?

Programming Exercises

1. Write a class of Rectangle, then derive a class of Cuboid, calculate the volume of the Cuboid object.

2. Write a class of two-dimensional point, then derive a class of three-dimensional point, calculate the distance between two three-dimensional points.

Part 4   File Operating

Chapter 11   Files and Stream

So far, all the data in our programs has been manually entered and retained in memory. Memory is volatile, that means when a program is terminated, the entire data is lost. Storing in a file will preserve your data even if the program terminates. 　保存
　易失的
　保存

If you have to enter a large number of data, it will take a lot of time to enter them all. However, if you have a file containing all the data, you can easily access the contents of the file using a few commands.

You can easily move your data from one computer to another without any changes.

In order to do this, we need to have a way of transferring data from variables in memory to a file on disk and from disk into memory. 　传输
　磁盘

11.1   Types of Files

When dealing with files, there are two types of files you should know about: text files and binary files. 　文本文件
　二进制文件

1. Text files
Text files are the normal .txt files. You can easily create text files using any simple text editors such as Notepad.

When you open those files, you'll see all the contents within the file as plain text. You can easily edit or delete the contents.

They take minimum effort to maintain, are easily readable, and provide the least security and take bigger storage space.

2. Binary files

Binary files are mostly the .bin files in your computer.

Instead of storing data in plain text, they store it in the binary form (0's and 1's).

They can hold higher amount of data, are not readable easily, and provides better security than text files.

In C and C++, you can perform four major operations on files, either text or binary:

- Creating a new file;
- Opening an existing file;
- Reading from and writing information to a file;
- Closing a file.

Though the mainly operations are similar, the commands are different, so we introduce separately.

11.2　File Operations in C++

In C++, files are mainly dealt by using three classes fstream, ifstream, ofstream. They are available in fstream header file. See Figure 11-1.　文件流；输入文件流；输出文件流

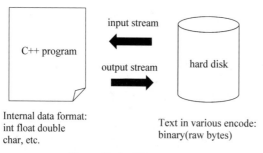

Figure 11-1　File storage

- ofstream: represents the output file stream and is used to create files and to write information to files.
- ifstream: represents the input file stream and is used to read information from

files.
- fstream: represents the file stream has the capabilities of both ofstream and ifstream which means it can create files, write information to files, and read information from files.

11.2.1 Creating/Opening a File Using Fstream

采用文件流创建/打开文件

Before you do any work with files, you have to include the following header:

```
#include<fstream>
```

This lets you declare variables or objects of type fstream. Like string, fstream is not a basic data type but a class. This allows us to call its functions and use its operators. The class fstream is one of a family of classes known as I/O streams. The concept of reading and writing to streams is not new, only the source and destination are different. The source of cin is the console which is also the destination of cout. For fstream, the source and destination are files on disk.

Before you start doing anything with a file, you have to open it. The open function in fstream takes two arguments, they are:

- The name of the file you are dealing with;
- What operation you want to perform (for now, just reading and writing indicated by the constants ios::in and ios::out respectively). The code below shows three steps of opening a file.

```
#include<iostream>
#include<fstream>
using namespace std;
int main()
{
    fstream fs; // Step 1: Creating object of fstream class
    fs.open("testfile.txt",ios::out); // Step 2: Creating new file
    if(!fs)         // Step 3: Checking whether file exist
    {
        cout<<"File creation failed";
    }
    else
    {
        cout<<"New file created";
```

```
    fs.close();        // Step 4: Closing file
  }
  return 0;
}
```

11.2.2 Writing Text Files

写文本文件

The easiest file I/O operation is writing to a text file. When we say a text file, we mean one where the bytes are displayed as characters that we can immediately read. This is as opposed to a binary file where the computer can immediately read it. The code below shows how to write "hello" to the file "testfile.txt".

```
#include<iostream>
#include<fstream>
using namespace std;
int main()
{
  fstream fs; // Step 1: Creating object of fstream class
  fs.open("testfile.txt",ios::out); // Step 2: Creating new file
  if(!fs) // Step 3: Checking whether file exist
  {
    cout<<"File creation failed";
  }
  else
  {
    cout<<"New file created";
    fs<<"Hello";      //Step 4: Writing to file
    fs.close();       //Step 5: Closing file
  }
  return 0;
}
```

If you're opening a file to write but it does not exist, it will be created and then opened (provided the location you are writing to is writable). If the file you are writing to does exist (and is not read only), it will be written from the start of the file over the top of anything that may have been there before. Depending on how fstream has been implemented (this may differ platform to platform and compiler to compiler), it may also truncate the file length to zero, i.e. it wipes everything in the original file. For this reason, it's quite a good idea to be careful when performing file operations.

Before you do any files access, it is worth checking that the file is indeed open. To do this call the is_open function on your fstream and make sure it returns true. Providing it does, you can then treat your fstream just as you would treat cout.

When you finish writing to your file, remember to call the close function otherwise the operating system may still think the file is in use.

11.2.3 Reading Text Files

读文本文件

Reading from a text file is slightly more difficult. Firstly, you have to split up what you've read in then allocate it to variables in your program. Secondly, you have lots of "invisible" control characters (tabs, newlines, etc.) That are waiting to trip you up.

绊倒你

```
#include<iostream>
#include<fstream>
using namespace std;
int main()
{
    fstream fs; // step 1: Creating object of fstream class
    fs.open("testfile.txt",ios::in); //Step 2: Creating new file
    if(!fs)     // Step 3: Checking whether file exist
    {
        cout<<"No such file";
    }
    else
    {
        char ch;
        while (!fs.eof())
        {
            fs >> ch;        // Step 4: Reading from file
            cout << ch;      // Message Read from file
        }
        fs.close();          // Step 5: Closing file
    }
    return 0;
}
```

Aside from the actual reading operations, the general form of reading a text file is quite similar to that of writing one. Just remember the following points:

- The open function requires ios::in as its second argument;
- There is no need to call flush prior to closing the file.

第二个参数

flush 函数，用于清空缓存

Once we have the file open for reading, the rules are pretty much the same as what we've been used to with cin: if you're reading a string, the resulting string will be filled with all the characters up to and including the first occurrence of whitespace. If you want text after that, you'll need to load it into another instance of string. Numeric variables can be read in the same way as you've already done with cin.

目标字符串（变量）

数值型变量

To get rid of some white space (or read from the stream a character at a time), there is a function you can use in fstream called get that extracts a single character from the stream. This could allow you to "eat" a character (such as a space or a tab) that you didn't want to find stuck on the front of the next string you were going to read.

提取

粘接在

There will come a point where you will run out of file to read. Nothing catastrophic happens if you just carry on reading using the ">>" operator, you just end up reading strings with nothing in them – these don't get assigned to string instance on the right of the ">>" operator, its value remains as it was. Luckily, the stream has an indicator inside it that can alert you if the end of the stream is reached. This is accessible through the function eof( ) in fstream which will return a true value if there is nothing left to read from the file. For example:

灾难性的

指示器；警告

```
fstream fs;
fs.open("basic_text.txt",ios::in);
if(fs.is_open())
{
    string line;
    while(!fs.eof())
        fs >> line;
    fs.close();
}
```

This will carry on reading strings from the file specified while the end of the file is not reached.

This has been a very brief introduction to reading and writing text files. There are a number of additional functions in fstream that we have not covered–feel free to go away and learn about them and use them in your programs. However, using combinations of the handful of functions you have learned here, you should be able to read and write text files that are complex enough to be useful.

NB: if you want to store the filename as a string instance, you will need to call its c_str function as you pass it as an argument to the open function of fstream. For example:

```
fstream fs;
string sometext = "basic_text.txt";
fs.open(sometext.c_str(),ios::in);
```

11.2.4 String Streams

字符串流

There will be times when you're programming that you will want to see what value the variables in your program take. Most people can't easily make the conversion between the numeric byte value and the numeric value, so it would be good to have some kind of conversion to human readable that was easy to use and required very little coding. Another common problem you will face is one of extracting multiple data items from a single piece of text. This lesson shows how both of these problems can be solved and applied in your programs with minimal effort.

提取多个数据项

Experience of using cout to print messages out to the command line should make you wonder: I can send any variable to cout and turn it into text, why can't I capture that text in a string? You might also wonder: if I can type in a number to the console, cin can convert it into a data type of my choosing – why can't I take text from anywhere and do the same? The good news is that the answer to both questions is: you can! The even better news is that if you know how to use cin and cout then you don't have to learn much more to do it. The solution lies in a set of classes called string streams. As the name suggests, these are streams (where data flows) that use a string instance as the source they read from and the destination they write to.

11.2.5 Converting to text

文本转换

As always, if you are using someone else's code, you need to provide the **appropriate headers**. For a string stream we need:

合适的头文件

```
#include<sstream>
```

There are two types of string stream. The first one we'll look at is ostringstream which is used to fill strings up with data. We can declare an instance of ostringstream called oss in the usual way:

```
ostringstream oss;
```

Next we'll push some data of different types into this string:

```
int year=2020;
int month=10;
int day=25;

oss << year;
oss << "-";
oss << month;
oss << "-";
oss << day;
```

Once we have all the data we want, we do the following to "close" the string (see what happens if you miss this out):

```
oss << ends;
```

You can then call the str() function in ostringstream to get the string you have created. If you want, you can then send it to the console to see what is in it:

```
string output=oss.str();
cout << output << endl;
```

And that's all you have to do–all of the complex details of converting data into strings are hidden away from you inside the ostringstream class. The whole program is:

```
#include<sstream>
```

```
#include<string>
#include<iostream>
using namespace std;
int main()
{
    ostringstream oss;
    int year=2020;
    int month=10;
    int day=25;

    oss << year;
    oss << "-";
    oss << month;
    oss << "-";
    oss << day;
    oss << ends;
    string output=oss.str();
    cout << output << endl;

    return 0;
}
```

The output is:

```
2020-10-25
```

Converting text

There will also come a time when you have data stored in a string. This could come from one of the following sources:

- User input from cin;
- A line read from a text file.

We would then want to break up this string instance to get the data represented by it. To do this, we use the other type of string stream: istringstream. This class allows us to take a string and extract data from it into variables. The header you use is the same one as you need for ostringstream:

```
#include<sstream>
```

Then creates an instance of the istringstream class called iss.

```
istringstream iss;
```

If we had some string instances that we wanted to read from such as:

```
string src= " 192 168 0 1";
```

We can load them one at a time into the istringstream for reading:

```
iss.str(src);
```

We can now use iss just like we use cin:

```
float num;
while(iss >> num)
    cout << "found: " << num << endl;
```

The ">>" operator extracts characters from the string that's loaded inside it until it finds white space. Then it converts this substring into data of the same type as the variable on the right of the ">>", which in the above case is a float. The ">>" operator returns true if there are more characters to be read from the string that was loaded. If there are no more characters and we want to break up another string into basic data types, we call the following function on our istringstream instance:

```
iss.clear();
```

The whole program is:

```
#include<iostream>
#include<sstream>
#include<string>
using namespace std;

int main()
{
    istringstream iss;
    string src="192 168 0 1";
    iss.str(src);
    float num;
    
    while(iss >> num)
        cout << "found: " << num << endl;
```

```
    iss.clear();

    return 0;
}
```

The output is:

```
found: 192
found: 168
found: 0
found: 1
```

11.3 File Operations in C

C语言的文件操作

When working with files in C, you need to declare a pointer of type file. This declaration is needed for communication between the file and the program.

```
FILE *fp;  // declare a pointer of type file
```

11.3.1 Opening a File

打开文件

Opening a file is performed using the fopen() function defined in the stdio.h header file. The syntax is:

```
fp = fopen("filname","mode");
```

For example:

```
fopen("C:\\cprogram\\testfile.txt","w");
```

If the file testfile.txt doesn't exist in the location c:\cprogram, the function fopen() will create a new file named testfile.txt and opens it for writing as per the mode "w".

The writing mode allows you to create and edit (overwrite) the contents of the file.

覆盖

If the file format is binary, the mode should be "wb":

```
fopen("C:\\cprogram\\testfile.txt","wb");
```

Table 11-1 shows the opening modes in standard I/O.

Table 11-1 Opening Modes in Standard I/O

Mode	Meaning of Mode	During Inexistence of file
r	Open for reading.	If the file does not exist, fopen() returns NULL.
rb	Open for reading in binary mode.	If the file does not exist, fopen() returns NULL.
w	Open for writing.	If the file exists, its contents are overwritten. If the file does not exist, it will be created.
wb	Open for writing in binary mode.	If the file exists, its contents are overwritten. If the file does not exist, it will be created.
r+	Open for both reading and writing.	If the file does not exist, fopen() returns NULL.
rb+	Open for both reading and writing in binary mode.	If the file does not exist, fopen() returns NULL.
a	Open for appending. Data is added to the end of the file.	If the file does not exist, it will be created.
ab	Open for appending in binary mode. Data is added to the end of the file.	If the file does not exist, it will be created.

Table 11-2 are the most important file management functions available in C:

Table 11-2 Most Important File Management Functions Available in C

Function	Purpose
fopen()	Creating a file or opening an existing file
fclose()	Closing a file
fprintf()	Writing a block of data to a file
fscanf()	Reading a block data from a file
getc()	Reading a single character from a file
putc()	Writing a single character to a file
getw()	Reading an integer from a file
putw()	Writing an integer to a file

11.3.2 Writing a File

写文件

After the file is opened, you can write data to it.

The code below shows how to write a integer number to a text file.

```
#include<stdio.h>
#include<stdlib.h>

int main()
{
   int num;
   FILE *fp;

   // use appropriate location
   fp = fopen("C:\\Ctest\\test.txt","w");

   if(fp == NULL)
   {
      printf("File open error!");
      exit(1);
   }

   printf("Enter a num: ");
   scanf("%d",&num);

   fprintf(fp,"%d",num);
   fclose(fp);

   return 0;
}
```

This program takes a number from the user and stores in the file test.txt.

After you compile and run this program, you can see a text file hamed test.txt created in the fold of Ctest on C drive of your computer. When you open the file, you can see the integer you have just entered.

11.3.3 Closing a File

关闭文件

After every successful file operations, a file need to be closed. Closing a file is performed using the fclose() function.

```
fclose(fp);
```

Here, fp is a file pointer associated with the file to be closed.

11.3.4 Read from a text file

You can read date from an existing file, the following code show how to do that:

```
#include<stdio.h>
#include<stdlib.h>

int main()
{
   int num;
   FILE *fp;

   // use appropriate location
   fp = fopen("C:\\Ctest\\test.txt","r");

   if(fp == NULL)
   {
      printf("File open error!");
      exit(1);
   }

   fscanf(fp,"%d", &num);
   printf("Value of num = %d", num);

   fclose(fp);
   return 0;
}
```

This program reads an integer stored in the test.txt file and prints it onto the screen.

If you successfully created the file from previous example, running this program will get you the integer you entered.

11.3.5 Writing Characters to a File

向文件里写字符

In C, when you write characters to a file, newline characters "\n" must be explicitly added. The stdio.h library offers the necessary functions to write to a file:

明确地

fputc(char, file_pointer): It writes a character to the file pointed to by file_pointer.

fputs(str, file_pointer): It writes a string to the file pointed to by file_pointer.

fprintf(file_pointer, str, variable_lists): It prints a string to the file pointed to by file_pointer. The string can optionally include format specifiers and a list of variables variable_lists.

The program below shows how to perform writing characters to a file:

```
#include<stdio.h>
#include<stdlib.h>
int main()
{
   FILE *fp;
   char str[] = "I love my university.\n";
   fp = fopen("c:\\testfile\\abc.txt", "w");
   if(fp == NULL)
   {
      printf("File open error!");
      exit(1);
   }
   for (int i = 0; str[i] != '\n'; i++)
   {
      /* write to file using fputc() function */
      fputc(str[i], fp);
   }
   fclose(fp);

   return 0;
}
```

The output is:

```
I love my university.
```

The program above takes each character of the array and writes it into the abc.txt until it reaches the next line symbol "\n" which indicates that the string reaches the end.

11.3.6 Appending Data to a File

追加数据到文件

When opening a file with "a" mode, the data can be added to the end of the file.

If the file does not exist, it will be created. The example code is as below:

```
fp = fopen("c:\\testfile\\abc.txt", "a");
```

In the example below, we have created and opened a file called abc.txt in appending mode. If the file is opened successfully, then write a string to the file using fputs() function. Then the file is closed using the fclose function.

```
#include<stdio.h>
#include<stdlib.h>
int main()
{
   FILE *fp;
   fp = fopen("c:\\testfile\\abc.txt", "a");
   if(fp == NULL)
   {
      printf("File open error!");
      exit(1);
   }
   fputs("\nI study hard in my university.", fp);
   fclose(fp);
   return 0;
}
```

If the previous example is successfully executed, the output should be:

```
I love my university.
I study hard in my university.
```

Chapter Review

This chapter will teach you the following useful programming skills:

- How to use fstream to read and write text files.
- What differences in function calls need to be made in order to open a file to: write and read.
- How do you know when you've run out of data to read from a file.
- One way of converting from any data type to a string.
- One way of converting from a string to any data type.
- How to use these to display the values your class instances have.

Programming Exercises

1. Write a program to write number 1 to 100 in a data file hamed notes.txt.

2. Write a program, which initializes a string variable to the content "Time is a great teacher but unfortunately it kills all its pupils. Berlioz" and outputs the string to the disk file hamed out.txt. you have to include all the header files if required.

3. Write a user-defined function to read the content from a text file hamed out.txt, count and display the number of alphabets present in it.

4. Write a function to count the number of blank present in a text file named "out.txt".

5. Write a function to count number of words in a text file named "out.txt".

Chapter 12   Splitting Program into Multiple Files

After more than two months of study, your programming ability improved dramatically, and the program you write is getting more and more complex, the code is getting longer and longer. It becomes more and more important to manage the code properly. 极大地

Of course, you could put thousands of lines of code in one file, but it is not reasonable. It will cause great difficulties for understanding, modifying, debugging and maintaining the code. 纠错

Further more, as your program grows, if everything is in a single file, then every line of code must be fully recompiled every time you make any little change. This might not seem like a big deal for small programs, but when you have a reasonable size project, compiling times can take several minutes, even minor change of the code, you have to wait that long time. 微小的

So, separating a large program into serval parts and organizing the code properly is becoming not only necessary but also important.

12.1   Separate a Program into Multiple Files in C   把一个程序文件分拆为多个文件

For a C program, a header file is a file with extension .h which contains C function declarations and macro definitions to be shared between several source files. There are two types of header files: the files that come with the compiler and the files that the programmer writes.

We have already used header files when we first used the printf( ) function, or other I/O functions by including the stdio.h file:

```
#include<stdio.h>
```

The stdio.h file is the header file that the compiler written for us, #include is a preprocessor directive.

The preprocessor goes and looks up the stdio.h file in the standard library.

A header file in standard library will be surrounded by a pair of angle brackets.

To include your own header files, quotes need to be used, like this:

`#include "myheaderfile.h"`

Including a header file is equal to copying the content of the header file to your program, but we do not do it because it will be error-prone and it is not a good idea to copy the content of a header file in the source files, especially if we have multiple source files in a program.

A simple practice in C or C++ programs is that we keep all the constants, macros, system wide global variables, and function prototypes in the header files and include that header file wherever it is required. Let's make an example. This program calculates the area of a triangle by using a function triangleArea. If we put everything in a single file, the program should be as follows:

```c
#include<stdio.h>
#include<math.h>

double triangleArea(double a, double b, double c)
{
    double p, s;
    p = (a+b+c)/2.0;
    s = sqrt(p*(p-a)*(p-b)*(p-c));
    return s;
}

int main()
{
    double a,b,c,area;
    printf("Please input the length of the triangle:");
    scanf("%lf%lf%lf",&a,&b,&c);
    area = triangleArea(a, b, c);
    printf("The area of triangle is:%lf",area);
    return 0;
}
```

Suppose that we want to separate this program into two parts: a header file and a .c file.

The name of the header file could be triangleArea.h. It includes the function prototype:

```c
// the triangleArea.h file
double triangleArea(double a, double b, double c);
```

Then create a .c file named triangleArea.c, put the function definition in this file.

```c
// the triangleArea.c file
#include "math.h"
#include "triangleArea.h"
double triangleArea(double a, double b, double c)
{
    double p, s;
    p = (a+b+c)/2.0;
    s = sqrt(p*(p-a)*(p-b)*(p-c));
    return s;
}
```

The main .c file just include the triangleArea.h file, which will make the triangleArea() function available for the main file:

```c
// the triangle.c file including main() function
#include<stdio.h>
#include "triangleArea.h"
int main()
{
    double a,b,c,area;
    printf("Please input the length of the triangle:");
    scanf("%lf%lf%lf",&a,&b,&c);
    area = triangleArea(a, b, c);
    printf("The area of triangle is:%lf\n",area);
    return 0;
}
```

12.2 Practice in Microsoft Visual C++ 6

We can try the program above on Microsoft visual C++ 6.

Step 1: Create a win32 console application named headerfilePratice (see Figure 12-1):

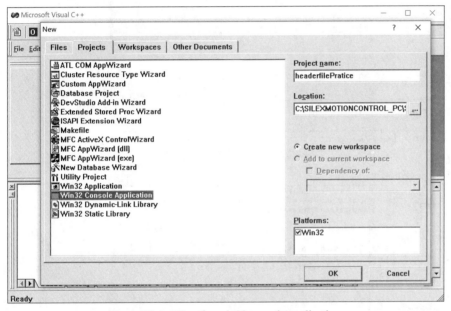

Figure 12-1   Create a win32 console application

Step 2: Select File → New, then select C/C++ Header File, choose the file name as: triangleArea.

Then copy the code of triangle Area.h in chapter 12.1. See Figure 12-2.

Step 3: Select File → New, select C/C++ Source File, the name is triangleArea, this file will hold the definition of the function. Then copy all the code in file triangleArea in chapter 12.1. See Figure 12-3.

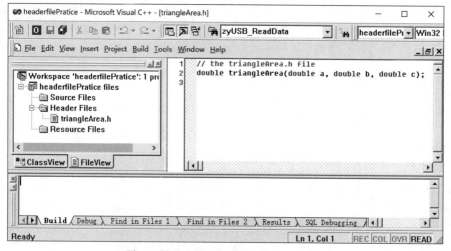

Figure 12-2　Creating triangleArea.h file

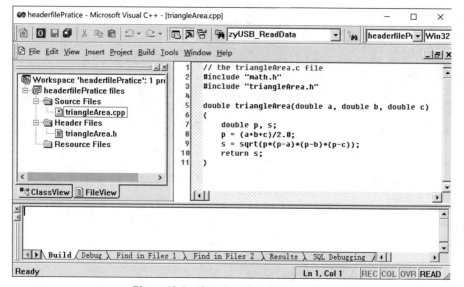

Figure 12-3　Creating triangleArea.c file

Step 4: Select File → New, select C/C++ Source File, the name is triangle, this file will include the main function. Then put all the code of file triangle.c in chapter 12.1. See Figure 12-4.

Chapter 12　Splitting Program into Multiple Files

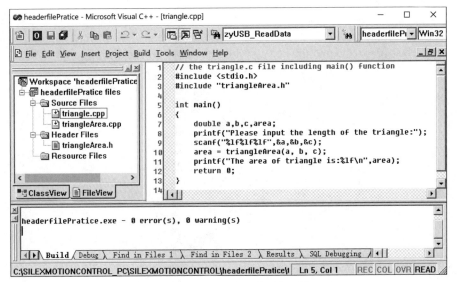

Figure 12-4　Creating triangle.c file

Then you can try to compile and execute the code.

12.3　Separate a Program into Multiple Files in C++

All of the classes that we have written so far have been simple enough that we have been able to implement the member functions directly inside the class definition itself.

However, as classes get longer and more complicated, having all the member function definitions inside the class can make the class harder to manage and work with. Using an already-written class only requires understanding its public interface (the public member functions), not how the class works underneath the hood.

外罩

Fortunately, C++ provides a way to separate the "declaration" portion of the class from the "implementation" portion. This is done by defining the class member functions outside of the class definition. To do so, simply define the member functions of the class as if they were normal functions, but prefix the class name to the function using the scope resolution operator (::) (same as for a namespace).

Take the air_ticket class as an example.

The declarations of base class air_ticket and the derived class business_air_ticket can be put into a header file, give it a name air_ticket.h. The code is as follows:

```cpp
#include<iostream>
#include<string>
#include<sstream>
using namespace std;

class air_ticket      // base class declaration
{
protected:
   string m_passenger;
   int m_flightNumber;
   string m_date;
   string m_origin;
   string m_destination;
   string m_carrier;
public:
   air_ticket();
   air_ticket(string,string);
   ~air_ticket();
   void set_passenger(string name);
   string get_passenger();
   void set_origin(string);
   void set_destination(string);
   string get_origin();
   string get_destination();
   string get_date_of_travel();
   string get_carrier();
   int get_flight_number();
   void set_flight_number(int);
   virtual bool lounge_access();
   virtual int get_seat_pitch();
   string print_ticket();
};

class business_air_ticket: public air_ticket  // derived class
                                              // declaration
{
```

```
private:
    int m_frequentFlyerPoints;
public:
    business_air_ticket(string,string);
    ~business_air_ticket();
    bool lounge_access();
    int get_seat_pitch();
};
```

The definitions of base class air_ticket and the derived class business_air_ticket are put into a cpp file, give it a name air_ticket.cpp. The code is as below:

```
#include "air_ticket.h"
// The base class definition
air_ticket::air_ticket()
{
    m_date = "23rd May 2015";
    m_carrier = "KLM";
}
air_ticket::air_ticket(string date,string airline)
{
    this->m_date=date;
    this->m_carrier=airline;
}
air_ticket::~air_ticket()
{
    //nothing to clean up
}
void air_ticket::set_origin(string org)
{
    this->m_origin=org;
}
void air_ticket::set_destination(string dest)
{
    this->m_destination=dest;
}
void air_ticket::set_passenger(string name)
{
    m_passenger = name;
```

```cpp
}
string air_ticket::get_passenger()
{
    return m_passenger;
}
string air_ticket::get_origin()
{
    return this->m_origin;
}
string air_ticket::get_destination()
{
    return this->m_destination;
}
string air_ticket::get_date_of_travel()
{
    return this->m_date;
}
string air_ticket::get_carrier()
{
    return this->m_carrier;
}
void air_ticket::set_flight_number(int num)
{
    this->m_flightNumber=num;
}
int air_ticket::get_flight_number()
{
    return this->m_flightNumber;
}
bool air_ticket::lounge_access()
{
    return false;
}

int air_ticket::get_seat_pitch()
{
    return 98;
}

string air_ticket::print_ticket()
{
```

```
    ostringstream oss;

    oss << endl;
    oss << "==============================" << endl;
    oss << "Name: " << this->m_passenger << endl;
    oss << "Flight: " << this->m_carrier << " " << this->
m_flightNumber;
    oss << endl;
    oss << endl;
    oss << "From: " << this->m_origin << endl;
    oss << "To: " << this->m_destination << endl;
    oss << endl;

    oss << "Lounge: ";

    if(this->lounge_access()==true)
        oss << "Y";
    else
        oss << "N";

    oss << " | Seat Pitch: " <<  get_seat_pitch() << " cm" << endl;
    oss << "Date: " << this->m_date << endl;
    oss << "==============================" << endl;
    oss << ends;

    return oss.str();
}
business_air_ticket::business_air_ticket(string date,string carr)
{
    string m_date = date;
    string m_carrier = carr;
}
// The derived class definition
business_air_ticket:: ~business_air_ticket()
{

}

bool business_air_ticket::lounge_access()
{
```

```
    return true;
}

int business_air_ticket::get_seat_pitch()
{
    return 155;
}
```

In main function, we just include the "air_ticket.h" file at the beginning. The program can work properly. The code is as below:

```
#include "air_ticket.h"

int main()
{
    air_ticket at1;
    at1.set_passenger("CZM");
    at1.set_origin("Shangahi");
    at1.set_destination("Beijing");
    at1.set_flight_number(999888);
    cout << at1.print_ticket() << endl;

    string details;
    business_air_ticket bat("20th September 2007","KLM");
    bat.set_passenger("Bruce");
    bat.set_origin("Amsterdam Schiphol");
    bat.set_destination("Helsinki Vantaa");
    bat.set_flight_number(1171);
    details=bat.print_ticket();

    cout << details << endl;

    return 0;
}
```

Then you can compile the code and run it.

Programming Exercises

1. For the program exercise 1 in chapter 10, create a win32 console application, and then splitting the program into three files: one header file for all the class declarations, one for corresponding definition file, and one file for main( ) function.

2. For the programming exercise 2 in chapter 10, create a win32 console application, and then splitting the program into three files: one header file for all the class declarations, one for corresponding definition file, and one file for main( ) function.

Part 5 Projects for C/C++

Part A Elementary C/C++ Projects

Project 1 Logical Gates   逻辑门

A Logic gate is an elementary building block of any digital circuits. It takes one, two or more inputs and produces output based on those inputs. Outputs may be high (1) or low (0). Logic gates are implemented using diodes or transistors. In a computer, most of the electronic circuits are made up logic gates. Logic gates are used to create a circuit that performs calculations, data storage or shows off the power of object-oriented programming especially the power of inheritance.

There are seven basic logic gates defined, they are:

- AND gate;   与门
- OR gate;   或门
- NOT gate;   非门
- NAND gate;   与非门
- NOR gate;   或非门
- XOR gate;   异或门
- XNOR gate.   异或非门

Below are the brief details about them along with their implementation:

Inputs and outputs can exist in two possible states:

a voltage level that represents On or True;

a second voltage level that represents Off or False.

The relationship between the inputs and the outputs follows the laws of Boolean algebra. The following logical functions exist for Digital Logic Gates.

1. AND gate(see Table A-1): output is true if all inputs are true, output is false if any input is false.

Table A-1   AND gate

Logical Function	Logical Symbol	Boolean Expression	Truth Table		
			inputs		output
			A	B	Y
AND	(A, B → Y)	$Y = A \cdot B$	0	0	0
			1	0	0
			0	1	0
			1	1	1

2. OR gate(see Table A-2): output is true if any or all inputs are true; output is false if all inputs are false.

Table A-2   OR gate

Logical Function	Logical Symbol	Boolean Expression	Truth Table		
			inputs		output
			A	B	Y
OR	(A, B → Y)	$Y = A + B$	0	0	0
			1	0	1
			0	1	1
			1	1	1

3. NOT gate (see Table A-3): output is true if input is false; output is false if input is true.

Table A-3   NOT gate

Logical Function	Logical Symbol	Boolean Expression	Truth Table	
			inputs	output
			A	Y
NOT	(A → Y)	$Y = \overline{A}$	0	1
			1	0

4. NAND gate (Inverted AND gate NOT AND) (see Table A-4): output is false if all inputs are true; output is true if any input is false.

Table A-4 NAND gate

Logical Function	Logical Symbol	Boolean Expression	Truth Table		
			inputs		output
			A	B	Y
NAND	A, B →⌐D⟩– Y	$Y = \overline{A + B}$	0	0	1
			1	0	1
			0	1	1
			1	1	0

5. XOR gate (see Table A-5): output is true if an odd number of inputs are true; output is false if an even number of inputs are true.

Table A-5 XOR gate

Logical Function	Logical Symbol	Boolean Expression	Truth Table		
			inputs		output
			A	B	Y
XOR	A, B →⟩D⟩– Y	$Y = A \oplus B$	0	0	0
			1	0	1
			0	1	1
			1	1	0

6. NOR gate (Inverted OR gate NOT OR) (see Table A-6): output is false if any or all inputs are true; output is true if all inputs are false.

Table A-6 NOR gate

Logical Function	Logical Symbol	Boolean Expression	Truth Table		
			inputs		output
			A	B	Y
NOR	A, B →D⟩– Y	$Y = \overline{A \cdot B}$	0	0	1
			1	0	0
			0	1	0
			1	1	0

7. XNOR gate (inverted XOR gate NOT XOR) (see Table A-7): output is true if

an even number of inputs are true; output is false if an odd number of inputs are true.

Table A-7 XNOR gate

Logical Function	Logical Symbol	Boolean Expression	Truth Table		
			inputs		output
			A	B	Y
XNOR		$Y = \overline{A} \oplus \overline{B}$	0	0	1
			1	0	0
			0	1	0
			1	1	1

In this project, you will create classes to represent some or all of the types of logic gates above. When creating a Gate object, the user should be able to specify the number of inputs to the gate in the range of 1 to 4. Getter and setter functions employing an index should be part of the class allowing the status of the gate inputs to be observed or changed. When any input is changed, the gate object must update its output appropriately. A getter function should be provided to allow the status of the output to be determined.

This project will benefit if the principles of inheritance are used with the development of a basic class dealing with setting the inputs and getting the values of the inputs and outputs. Classes to implement specific logic gates can be created as descendants of this base class.

Note C++ contains the logical operators && (AND)、|| (OR) and ! (NOT) which can be used with Boolean variables.

Figure A-1 is UML Diagram for AND_gate class.

Class declaration for 2 input AND_gate class is as below:

```
class and_gate
{
private:
    bool input1;
    bool input2;
    int no_inputs;
```

Part A Elementary C/C++ Projects

AND_gate
- bool input1
- bool input2
- int no_inputs
- bool output
+ and_gate();
+ void set_input_1(bool)
+ void set_input_2(bool)
+ void set_output()
+ bool get_input_1()
+ bool get_input_2()
+ bool get_output()

Figure A-1   UML Diagram for AND_gate class

```
   bool output;
public:
   and_gate();
   void set_input_1(bool);
   void set_input_2(bool);
   void set_output();
   bool get_input_1();
   bool get_input_2();
   bool get_output();
};
```

Definition for some member functions are as below:

```
and_gate::and_gate()// constructor
{
  this->input1 = false;
  this->input2 = false;
  this->output = false;
}
void and_gate::set_input_1(bool input)
{
  this->input1 = input;
  this->set_output();
}
void and_gate::set_output()
```

```
{
    this->output = this->input1 && this->input2;
}
bool and_gate::get_input_1()
{
    return this->input1;
}
```

You are required to complete the entire program.

Project 2 Lorry Fleet

A program is required to hold data on a fleet of vehicles used by a delivery firm. The firm has 10 vehicles, each vehicle can carry a load of 10000kg. Design a class which represents one of the vehicles used by the company. The class should contain member variables for:

(1) the lorry's number plate.

(2) the amount of cargo it is carrying.

(3) a description of the cargo.

(4) where it is taking the cargo from.

(5) where it is delivering the cargo to.

The following member functions should be included in the class:

(1) A getter function to allow the value of the number plate to be read.

(2) Getter and setter member functions that allow the contents of the following member variables to be read or changed:

- the amount of cargo;
- the description of the cargo;
- the source of the cargo;
- the destination of the cargo.

A second class should be designed that contains a member array that will hold 10 objects of the vehicle type. Methods should be provided in this class which are passed an index to the array. These methods should allow the value of the number plate of a lorry in the fleet to be read, the quantity of cargo, description of the cargo, source and destination of a lorry in the fleet to be read or to be changed.

Figure A-2 is UML Diagram for lorry class and Figure A-3 is UML Diagram for fleet class.

Lorry
- string number_plate
- int cargo_weight
- string cargo_info
- string source
- string destination
- int max_cargo

+ lorry()
+ lorry(string)
+ string get_cargo_description
+ string get_cargo_discription()
+ string get_destination()
+ string get_source()
+ void set_cargo_weight(int)
+ void set_cargo_description(string)
+ void set_destination(string)
+ void set_source(string)

Figure A-2　UML Diagram for lorry class

Fleet
- lorry* fleet_array
- int size
+ fleet()
+ ~fleet()
+ string get_lorry_numberplate(int)
+ int get_lorry_cargo_weight(int)
+ string get_lorry_cargo_description(int)
+ string get_lorry_destination(int)
+ string get_lorry_source(int)
+ void set_lorry_cargo_weight(int,int)
+ void set_lorry_cargo_description(string,int)
+ void set_lorry_destination(string,int)
+ void set_lorry_source(string,int)

Figure A-3　UML Diagram for fleet class

Code for lorry class declaration is as below:

```
class lorry
{
public:
    lorry();        //default constructor
    lorry(string);//constructor with number plate
    string get_numberplate(); //getter member function for
                              //numberplate
    int get_cargo_weight();
    string get_cargo_discription();
    string get_destination();
    string get_source();

    void set_cargo_weight(int);
    void set_cargo_description(string);
    void set_destination(string);
    void set_source(string);

private:
    string number_plate;
    int cargo_weight;
    string cargo_info;
    string source;
    string destination;
    int max_cargo;
};
```

Class definition for some member functions is as below:

```
lorry::lorry()
{
  this->number_plate="TFK 685";
  this->max_cargo=10000;
}
// default constructor
lorry::lorry(string number_plate)
{
  this->number_plate=number_plate;
  this->max_cargo=10000;
}
// constructor with number plate
string lorry::get_numberplate()
{
```

```
    return this->number_plate;
}
void lorry::set_cargo_weight(int weight)
{
    if (weight >= this->max_cargo)
    {
        this->cargo_weight = weight;
    }
    else
    {
        cout <<"cargo to large for lorry"<<endl;
    }
}
```

Code for fleet class declaration is as below:

```
class fleet
{
public:
    fleet();
    ~fleet();
    string get_lorry_numberplate(int);
    int get_lorry_cargo_weight(int);
    string get_lorry_cargo_description(int);
    string get_lorry_destination(int);
    string get_lorry_source(int);
    void set_lorry_cargo_weight(int,int);
    void set_lorry_cargo_description(string,int);
    void set_lorry_destination(string,int);
    void set_lorry_source(string,int);
private:
    lorry* fleet_array;
    int size;
};
```

Partial code listing for some member functions is as below:

```
fleet::fleet()
{
    this->size = 3;
```

```cpp
    this->fleet_array = new lorry[this->size];

    this->fleet_array[0]=lorry("LOR 000");
    this->fleet_array[1]=lorry("LOR 001");
    this->fleet_array[2]=lorry("LOR 002");
}
fleet::~fleet()
{
    delete [] fleet_array;
    fleet_array = NULL;
}
string fleet::get_lorry_numberplate(int index)
{
    return this->fleet_array[index].get_numberplate();
}

void fleet::set_lorry_cargo_weight(int weight, int index)
{
    this->fleet_array[index].set_cargo_weight(weight);
}
```

You are required to complete the entire project.

Project 3   Money Class

In China, money in the form of notes is available as follows: 100RMB; 50RMB; 20RMB; 10RMB; 5RMB and 1RMB.   纸币

In this project, you are to create a class that will model the contents of a person's wallet or purse. The class that you develop should contain member variables which contain the total number of notes of each value that the person has.

The class should provide member functions that allow the total amount of money the person has in their purse or wallet and the number of notes of each size that they have to be changed.

A Member functions should also be provided to allow money to be added to the class.

A final member function should allow the person to pay for something with their money. This function should update the number of notes of each denomination correctly to match the new total for the money that the person has available after their purchase. If it is possible the function should use the minimum number of notes to pay the correct amount of money for a purchase.

Figure A-4 is UML diagram for money class.

Code for money class declaration is as below:

```
class money
{
public:
    money();
    money(int);
    int get_total_money();
    int get_no_rmb100();
    int get_no_rmb50();
    int get_no_rmb20();
    int get_no_rmb10();
    int get_no_rmb5();
    int get_no_rmb1();
    void add_money(int);
```

Money
- int rmb_100
- int rmb_50
- int rmb_20
- int rmb_10
- int rmb_5
- int rmb_1
+ public
+ money()
+ money(int);
+ int get_total_money()
+ int get_no_rmb100()
+ int get_no_rmb50()
+ int get_no_rmb20()
+ int get_no_rmb10()
+ int get_no_rmb5()
+ int get_no_rmb1()
+ void add_money(int)
+ void spend_money(int)

Figure A-4 UML diagram for money class

```
    void spend_money(int);
private:
    int rmb_100;
    int rmb_50;
    int rmb_20;
    int rmb_10;
    int rmb_5;
    int rmb_1;
};
```

Class definition for some member functions is as below:

```
money::money()// constructor
{
    this->rmb_100 = 2;
    this->rmb_50 = 3;
    this->rmb_20 = 4;
    this->rmb_10 = 6;
    this->rmb_5 = 5;
```

```
        this->rmb_1 = 4;
}
int money::get_total_money()
{
    int total = 0;
    total = total + (this->rmb_100 *100);
    total = total + (this->rmb_50 *50);
    total = total + (this->rmb_20 *20);
    total = total + (this->rmb_10 *10);
    total = total + (this->rmb_5 *5);
    total = total+ this->rmb_1;
    return total;
}
int money::get_no_rmb100()
{
   return this->rmb_100;
}
```

You are required to complete the entire project.

Part B Computer Game and Machine Learning Projects

Project 4 Hangman Game Project   猜单词游戏

In the game of Hangman, the computer chooses a word at random from a given list of words. The player then guesses the word by one letter at a time. If the letter that the player guessed is in the answer, all occurrences of that letter are revealed to the player.   显示出来

The game ends when the player has guessed every letter in the word, before he reaches the allowed number of strikes (usually 5). This program is an interactive Hangman game.   尝试；交互的

This project asks the player to guess a city name of China. The city name is expressed in Chinese phonetic alphabet. In the demo code, the city names are stored in a one-dimensional character array. You are required: 汉语拼音 一维字符型数组

(1) Modify the code to store the city names in a text file.
(2) Add a scoring system, can give a score to every player. And record the highest score of all the players.   打分系统
(3) Additionally, a good UI is expected.   用户界面

Partial code listing for some functions is as below:

```
// Take one character (guess) and the secret word, and fill in the
// unfinished guess word. Returns number of characters matched.
// Also, returns zero if the character is already guessed.

int letterFilling (char guess, string secretword, string &guessword)
{
    int i;
```

```cpp
        int matches=0;
        int len=secretword.length();
        for (i = 0; i < len; i++)
        {
            // Did we already match this letter in a previous guess?
            if (guess == guessword[i])
                return 0;
            // Is the guess in the secret word?
            if (guess == secretword[i])
            {
                guessword[i] = guess;
                matches++;
            }
        }
        return matches;
}
#include<iostream>
#include<cstdlib>
#include<ctime>
#include<string>
using namespace std;

const int MAX_TRIES=5;

    system("color 5e");
    string name;
    char letter;
    int num_of_wrong_guesses=0;
    string word;
    string words[] =
    {
        "Jiangsu",
        "Shandong",
        "Zhejiang"
    };

    //choose and copy a word from array of words randomly
    srand(time(NULL));
    int n=rand()% 10;
    word=words[n];
```

```cpp
    // Initialize the secret word with the * character.
    string unknown(word.length(),'*');

    // welcome the user
    cout << "\n\nWelcome to hangman...Guess a country Name";
    cout << "\n\nEach letter is represented by a star.";
    cout << "\n\nYou have to type only one letter in one try";
    cout << "\n\nYou have " << MAX_TRIES << " tries to try and guess the word.";
    cout << "\n~~~~~~~~~~~~~~~~~~~~~~~~~~~~~~~~~~~~~~~~~~~";
    // Loop until the guesses are used up
    while (num_of_wrong_guesses < MAX_TRIES)
    {
        cout << "\n\n" << unknown;
        cout << "\n\nGuess a letter: ";
        cin >> letter;
        // Fill secret word with letter if the guess is correct,
        // otherwise increment the number of wrong guesses.
        if (letterFilling(letter, word, unknown)==0)
        {
            cout << endl << "Whoops! That letter isn't in there!" << endl;
            num_of_wrong_guesses++;
        }
        else
        {
            cout << endl << "You found a letter! Isn't that exciting!" << endl;
        }
        // Tell user how many guesses has left.
        cout << "You have " << MAX_TRIES - num_of_wrong_guesses;
        cout << " guesses left." << endl;
        // Check if user guessed the word.
        if (word==unknown)
        {
            cout << word << endl;
            cout << "Yeah! You got it!";
            break;
        }
    }
    if(num_of_wrong_guesses == MAX_TRIES)
```

```
    {
        cout << "\nSorry, you lose...you've been hanged." << endl;
        cout << "The word was : " << word << endl;
    }
    cin.ignore();
    cin.get();
```

Part B Computer Game and Machine Learning Projects

Project 5 The tic-tac-toe game

三连棋（叉圈游戏）

The tic-tac-toe game is played on a 3×3 grid by two players, who take turns. The first player marks moves with a cross, the second with a circle. The player who has formed a horizontal, vertical, or diagonal sequence of three marks wins. This program draws the game board, ask the players for the coordinates of the next mark, change the players after every successful move and display the winner.

3×3 格子

棋盘；坐标

The demo code is for two human players, you are required to:

(1) Modify the code, change one human player to computer controlled AI player.
(2) Add a scoring system, can give a score to every player. And record the highest score of all the players.
(3) Additionally, a good UI is expected.

```
#include<iostream>
using namespace std;

char square[10] = {'0','1','2','3','4','5','6','7','8','9'};
int main()
{
    int player=1,i,choice;
    char mark;
    do
    {
        board();
        player=(player%2)?1:2;
        cout << "Player " << player << ", enter a number: ";
        cin >> choice;
        mark=(player == 1) ? 'X' : 'O';

        if (choice == 1 && square[1] == '1')
            square[1] = mark;
        else if (choice == 2 && square[2] == '2')
            square[2] = mark;
        else if (choice == 3 && square[3] == '3')
            square[3] = mark;
        else if (choice == 4 && square[4] == '4')
            square[4] = mark;
```

205

```
            else if (choice == 5 && square[5] == '5')
                square[5] = mark;
            else if (choice == 6 && square[6] == '6')
                square[6] = mark;
            else if (choice == 7 && square[7] == '7')
                square[7] = mark;
            else if (choice == 8 && square[8] == '8')
                square[8] = mark;
            else if (choice == 9 && square[9] == '9')
                square[9] = mark;
            else
            {
                cout<<"Invalid move ";
                player--;
                cin.ignore();
                cin.get();
            }
            i=checkwin();
            player++;
    }while(i==-1);
    board();
    if(i==1)
        cout<<"==>\aPlayer "<<--player<<" win ";
    else
        cout<<"==>\aGame draw";
    cin.ignore();
    cin.get();
    return 0;
}

/*********************************************
    FUNCTION TO RETURN GAME STATUS
    1 FOR GAME IS OVER WITH RESULT
    -1 FOR GAME IS IN PROGRESS
    0 GAME IS OVER AND NO RESULT
*********************************************/

int checkwin()
{
    if (square[1] == square[2] && square[2] == square[3])
        return 1;
```

```cpp
    else if (square[4] == square[5] && square[5] == square[6])
        return 1;
    else if (square[7] == square[8] && square[8] == square[9])
        return 1;
    else if (square[1] == square[4] && square[4] == square[7])
        return 1;
    else if (square[2] == square[5] && square[5] == square[8])
        return 1;
    else if (square[3] == square[6] && square[6] == square[9])
        return 1;
    else if (square[1] == square[5] && square[5] == square[9])
        return 1;
    else if (square[3] == square[5] && square[5] == square[7])
        return 1;
    else if (square[1] != '1' && square[2] != '2' && square[3] != '3' && square[4] != '4' && square[5] != '5' && square[6] != '6' && square[7] != '7' && square[8] != '8' && square[9] != '9')
        return 0;
    else
        return -1;
}

// Function to draw board of Tic Tac with player mark
void board()
{
    system("cls");
    cout << "\n\n\tTic Tac Toe\n\n";

    cout << "Player 1 (X)  -  Player 2 (O)" << endl << endl;
    cout << endl;

    cout << "     |     |     " << endl;
    cout << "  " << square[1] << "  | " << " " << square[2] << "  | " << " " << square[3] << endl;

    cout << "_____|_____|_____" << endl;
    cout << "     |     |     " << endl;

    cout << "  " << square[4] << "  | " << " " << square[5] << "  | " << " " << square[6] << endl;
```

```
    cout << "____|____|____" << endl;
    cout << "    |    |    " << endl;

    cout << "  " << square[7] << " | " << square[8] << " | " << square[9] << endl;

    cout << "    |    |    " << endl << endl;
}
```

Part B Computer Game and Machine Learning Projects

Project 6 Designing a chatbot

聊天机器人

In this project, instructions are provided to reveal how to write a fun and simple chatterbot like application in C++. The user enters a phrase and then the learner responds if the initial phrase exists in memory. If it doesn't exist in memory, then the user can teach the learner what to say. The open source speech synthesizer "eSpeak" is used to create audible output. You can get it here:

展示

开源语音合成器
声音输出

http://sourceforge.net/projects/espeak/

The "eSpeak" files should be put in the same directory as the chatterbot application, if not, any audible message will not be heard.

There are five files in this project: two header files with corresponding cpp files, and one main cpp file with main() function in it.

The voice header file together with the cpp file will textually and audibly communicates a phrase. The open source eSpeak speech synthesizer is used to create the audible message. If the eSpeak exe is not located in the directory, no audible message will be heard.

eSpeak 可执行文件

The learner header file together with the cpp file will look for the passed phrase in the memory file. If there is a match, the associated response, stored below the initial phrase, will be outputted. If the response cannot be found, the learner will repeat the phrase, and prompt the user to enter an ideal response. This response will be stored in the memory file along with the initial phrase.

In the main cpp file, the main loop will continue until the application is closed. The user enters their input, and then the learner will respond.

The main.cpp file is as below:

```
#include<iostream>
#include "learner.h"
using namespace std;
main()
{
```

```
    Learner AI;    // Create a learner object
    for(;;)
    {
        cout << "\nYOU: ";    // User prompt
        string phrase;
        getline(cin, phrase);    // Using getline for multi word
                                 // input, then store in phrase
        cout << "COMPUTER: ";
        AI.respond(phrase);    // Pass the user input to the learner
                               // and see if he can respond
    }
}
```

The code for voice.h file is as below:

```
#include<iostream>

using namespace std;

class Voice {
public:
    void say(string phrase);    // Used to textually and audibly
                                // communicate a phrase
};
```

The code for voice.cpp file is as follow:

```
#include "voice.h"
#include<iostream>
#include<windows.h>

using namespace std;

/*
    The following function textually and audibly communicates a
phrase.
    The open source eSpeak speech synthesizer is used to create the
audible message.
    If the eSpeak exe is not located in the directory, no audible
message will be heard.
*/
void Voice::say(string phrase){
```

```
    string command = "espeak \"" + phrase + "\""; // Concat the
//phrase to the command
    const char* charCommand = command.c_str();   // Convert to a
//const char*
    cout << phrase << endl; // Textually output phrase
    system(charCommand);    // Send the command to cmd to execute
//espeak with the phrase an argument
}
```

The code for Learn.h file is as follows:

```
#include<iostream>
#include<fstream>
#include "voice.h"
using namespace std;
class Learner {
public:
    void respond(string phrase);  // Used to get, or teach a
                                  //response
    void say(string phrase);      // Used to textually and audibly
                                  //communicate a phrase
    Voice voice;   // The learner's voice that will audibly
                   //communicate a response
};
```

The code for Learn.cpp file is as follows:

```
#include "learner.h"
#include<iostream>
#include<fstream>

using namespace std;
void Learner::respond(string phrase)
{
    fstream memory;
    memory.open("memory/memory.txt", ios::in); // Open the memory
                                               //file for input

    // Search through the file until the end is reached
    while(!memory.eof()){          // While not at end of file
        string identifier;
        getline(memory,identifier);    // Get next phrase
```

```
        if(identifier == phrase){    // Is it the phrase we are  looking
                                     //for string response;
            getline(memory,response);   // If so, get the response
            voice.say(response);     // Textually and audibly output
                                     //the response!
            return;      // Leave the function
        }
    }

    memory.close();      // Looks like we couldn't find the phrase
                         //in memory. Close the file!
    memory.open("memory/memory.txt", ios::out | ios::app);
    // Now open for output, and append at end of file
    memory << phrase << endl;    // Record initial phrase in memory
    voice.say(phrase);    // Repeat the phrase the user entered
                          //string userResponse;
    cout << "YOU: ";
    getline(cin, userResponse);       // Get the ideal response
    memory << userResponse << endl; // Write the ideal response
                                    //to memory
    memory.close();      // Close the file!
    }

void Learner::say(string phrase){
    this->voice.say(phrase);
}
```

You are required to complete the project.

Part C　Hardware Based Projects

Project 7　Blinking a LED and LED chaser

让 LED 闪烁和 LED 追逐灯

LEDs are small, powerful lights that are used in many different applications. To start, we will work on blinking an LED, which is the "Hello World" of microcontrollers. It is as simple as turning a light on and off. Establishing this important baseline will give you a solid foundation and open the door to the electronic world.

Step 1:　Hardware Required

- Arduino uno (with USB cable)　×1
- Breadboard　×1
- LED　×1
- 220Ω Resistor　×1
- Jumper cables

USB 电线
面包板

电阻
连接线

Step 2:　Circuit Connection

To light an external LED, connect one end of the resistor to the digital pin, then connect the long leg of the LED (the positive pin, called the anode) to the other end of the resistor. Connect the short leg of the LED (the negative pin, called the cathode) to the GND. In the diagram (see Figure C-1), we show an UNO board that has a LED and a resistor connected.

电路连接

正极；阳极
负极
阴极

The better way is to use a breadboard for circuit connection (see Figure C-2).

Components like resistors need to have their terminals bent into 90° angles in order to fit the breadboard sockets properly(see Figure C-3). You can also cut the terminals shorter (see Figure C-4).

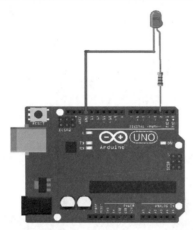

Figure C-1　Circuit diagram for blinking a LED

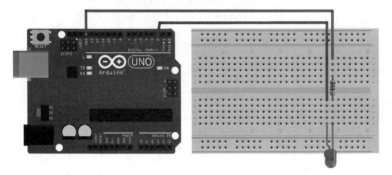

Figure C-2　Breadboard connection for blinking a LED

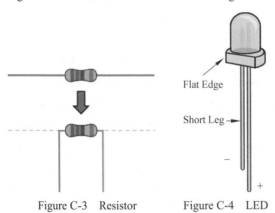

Figure C-3　Resistor　　　Figure C-4　LED

After the circuit is built, connect the Arduino board to your computer via a USB cable, then start the Arduino software (IDE) and begin to write code (see Figure C-5).

Part C Hardware Based Projects

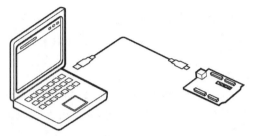

Figure C-5   Connecting a UNO board to a computer

Step 3:   Programming

This is where the real magic of Arduino happens. The Arduino board has a microcontroller on it which can be programmed to do almost anything you can think up. The code we need to write for this is fairly simple, but firstly you need to download the Arduino IDE (see Figure C-6) from their website (www.arduino.cc).

Figure C-6   The Arduino IDE

The Arduino IDE is a cross-platform Integrated Development Environment. This means that you can run it on every operating system. Once you have downloaded the IDE, follow the instructions on how to install it based on your operating system (Mac, Windows or Linux).

215

After launching the Arduino IDE, you should see a tab filled with two basic Arduino functions: the setup( ) and loop( ). These are built in library functions. The Arduino program is called "sketch".

The setup( ) function is called when a sketch starts. Use it to initialize the variables, pin modes, start using libraries, etc. The setup function will only run once, after each power up or reset of the Arduino board. Type the code below:

```
//Turns on an LED on for one second, then off for one second,
//repeatedly.

// the setup function runs once when you press reset or power the
//board

void setup() // initialize digital pin 2 as an output.
{
    pinMode(2, OUTPUT);
}

// the loop function runs over and over again forever
void loop()
{
    digitalWrite(2, HIGH); // turn the LED on (HIGH is the voltage
                           //level)
    delay(1000); // wait for a second
    digitalWrite(2, LOW); // turn the LED off by making the voltage
                          //LOW
    delay(1000); // wait for a second
}
```

After creating a setup() function, which initializes and sets the initial values, the loop() function does precisely what its name suggests, it loops consecutively, 连续地 allowing your program to change and respond. Use it to actively control the Arduino board. Let's look at the code in detail:

pinMode(2, OUTPUT)—Before you can use one of Arduino's pins, you need to tell Arduino Uno whether it is an input or output. We use a built-in "function" called pinMode() to do this.

digitalWrite(2, HIGH)—When you are using a pin as an output, you can command it to be high (output 5 volts), or low (output 0 volts).

Going Further—control more LEDs

As we have already known how to control one LED, it is not difficultly to control more LEDs. Then we can write code to control more LEDs to be lighted in different patterns. One of these patterns is called LED chaser. The LED chaser is one of the most popular types of LED-driving circuit and is widely used in advertising displays and in running-light "rope" displays in small discos, etc.

In the example below, we use ten LEDs to display in different patterns.

```
int i;
void setup() { // initialize digital pin 2-11 as an output.
    for (i = 2; i < 11; i++)
    {
        pinMode(i, OUTPUT);
    }
}
void loop() {
    for (i = 2; i <= 11; i++)
    {
        digitalWrite(i, HIGH);
        delay(100);
    }
    for (i = 11; i > 1; i--)
    {
        digitalWrite(i, LOW);
        delay(100);
    }
}
```

You are required to modify the code, and light these LEDs in your patterns.

Project 8 Distance Measurement using Ultrasonic Sensor and Arduino

An ultrasonic sensor emits an ultrasound at 40kHz that travels through the air, and if there is an object or obstacle in its path, it will bounce back to the module (see Figure C-7). The distance can be calculated by the travel time and the speed of the sound.

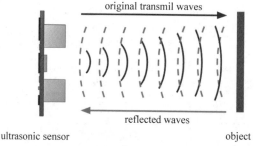

Figure C-7 Principle of measuring distance using a ultrasonic sensor

At 20°C, the speed of sound is roughly 343 m/s. Multiply the speed of sound by the time the sound waves traveled, you get the distance that the sound waves traveled.

$$\text{Distance (cm)} = \text{Speed of sound (cm/µs)} \times \text{Time (µs)} / 2$$

The speed of sound actually depends strongly on temperature and the humidity of the air. It is states that the speed of sound increases with roughly 0.6 m/s per degree Celsius. For most cases at 20°C, you can just use 343 m/s. But if you want to get more accurate readings, you can calculate the speed of sound with the following formula:

$$V \text{ (m/s)} = 331.3 + (0.606 \times T)$$
$$V = \text{speed of sound (m/s)}$$
$$T = \text{air temperature (°C)}$$

How the HC-SR04 Works

In this project, the type of Ultrasonic sensor is HC-SR04.

At the front of the HC-SR04 sensor, you can find two silver cylinders (ultrasonic transducers), one is the transmitter of the sound waves and the other is the receiver. To let the sensor generates a sonic burst, you need to set the trig pin high for at least 10 μs. The sensor then creates an 8-cycle burst of ultrasound at 40kHz.

超声波换能器
发射器
一段；触发

This sonic burst travels at the speed of sound and bounces back and gets received by the receiver of the sensor. The echo pin then outputs the time that the sound waves traveled in microseconds.

You can use the pulseIn( ) function in the Arduino code to read the length of the pulse from the echo pin. After that, you can use the formula mentioned above to calculate the distance between the sensor and the object.

Figure C-8 shows the diagram of the circuit.

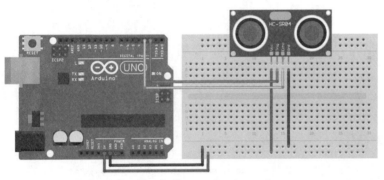

Figure C-8   Circuit diagram for ultrasonic sensor

The code below shows how to measure the distance using ultrasonic sensor.

```
// Defining Trig and Echo pin:
#define trigPin 2
#define echoPin 3

// Declaring variables:
long duration;
int distance;

void setup()
{
    pinMode(trigPin, OUTPUT);    // Defining input
```

```
    pinMode(echoPin, INPUT);    // Defining output
    Serial.begin(9600);    //Serial communication
}

void loop()
{
    digitalWrite(trigPin, LOW); // Clear the trigPin by setting it LOW
    delayMicroseconds(5);
    // Trigger the sensor by setting the trigPin high for 10
    //microseconds:
    digitalWrite(trigPin, HIGH);
    delayMicroseconds(10);
    digitalWrite(trigPin, LOW);
    // Read the echoPin, pulseIn() returns the duration (length of
    //the pulse) in microseconds:
    duration = pulseIn(echoPin, HIGH);
    distance = duration * 0.034 / 2; // Calculating the distance
    // Print the distance on the Serial Monitor (Ctrl+Shift+M):
    Serial.print("Distance = ");
    Serial.print(distance);
    Serial.println(" cm");
    delay(50);
}
```

Going Further—Intelligent Car Reverse Parking System

As we have already known how to measure the distance using a ultrasonic sensor, we can do a lot of things based on that project. In this project, an intelligent car reverse parking system (see Figure C-9) is built. A small ultrasonic sensor will calculate the distance of the car from the object behind. If the distance decreases to a specific range, the green LED, yellow LED and red LED will light, and a buzzer will beep in different frequency, let the driver know when to stop. 蜂鸣器

You are required to complete the project.

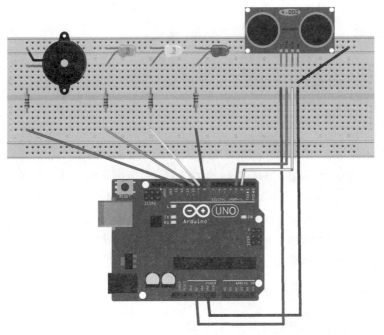

Figure C-9 Circuit diagram for car reverse parking system

Project 9　Servo Motor Projects

Servo motors can be found in robotic arms, cameras, CNC machines, printing presses, and other engineering applications where precision and repeated movement are required.

There are many types of servo motors in the market. In this project, a low-cost, small size servo motor (see Figure C-10) is used. It is essential a DC motor with a potentiometer mounted on the rotating plane as a feedback component. It contains a series of gears to slow down the rotation and increased the torque, and make the movement smoothly. A servo motor uses a circuit to control and send feedback information to a given controller.

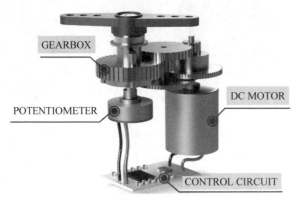

Figure C-10　Inner structure of a servo motor

A servo motor (see Figure C-11) traditionally has three wires: power, ground, and signal. The power cable is usually red, and in this configuration is connected to 5V. The ground wire is usually black or brown and must be connected to the ground. The signal cable is usually white, yellow, or orange and needs to be connected to the PWM (pulse width modulation) signal. For generating a PWM waveform a microcontroller is often used, because it has specific hardware to perform this task. It can do this with just a few lines of code.

The PWM is used for servo control. For PWM wave, duty cycle is used to control the angle of the servo rotation.

Figure C-11 Servo motor with wire

Duty cycle is referred to the percentage of time when the signal is high. The duty cycle specifically describes the percentage of time a digital signal is on over an interval or period of time. This period is the inverse of the frequency of the waveform (see Figure C-12).

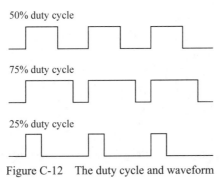

Figure C-12 The duty cycle and waveform

If a digital signal spends half of the time on and the other half off, we would say that the digital signal has a duty cycle of 50% and resembles an ideal square wave. If the percentage is higher than 50%, the digital signal spends more time in the high state than the low state and vice versa if the duty cycle is less than 50%.

100% duty cycle would be the same as setting the voltage to 5 Volts (high). 0% duty cycle would be the same as grounding the signal.

This system generates a square wave which changes the amount of time when the pulse is high, keeping the same period; a high level duration of the signal indicates the position where we want to put the motor shaft. An integrated potentiometer in the control circuitry monitors the current angle of the servo: if the axis is at a right angle, then the motor is off. The circuit checks that if the angle is not correct, the servo will correct the direction until the angle is right. Normally, the axis of the servo is able to reach around 180 degrees, but in some servos it reaches 210 degrees or even 360 degrees. The pulse duration indicates the rotation angle of the motor (see Figure C-13).

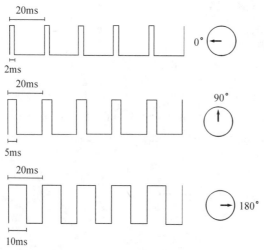

Figure C-13   The duty cycle and rotation angle

What are the functions in servo library?

To use a servo, a "Servo" header file needs to be included, then you can create a servo object:

```
Servo servo_1;
```

servo_1:   the name that you'll use when calling the servo to do something.

There are two member functions often used to control a servo in servo library; one is attach, and the other is write. The attach function is used to attach the servo to a pin. The syntax is as below:

```
servo_1.attach(pin_number);
```

pin_number is the pin of the UNO bord to attach servo.

The write function is used to set the position of the axis of the servo. The syntax is as below:

servo_1.write(pos);

pos = the angle that you want your servo motor to go. (it has to be something between 0° and 180°. Figure C-14 shows the circuit diagram for a servo motor.

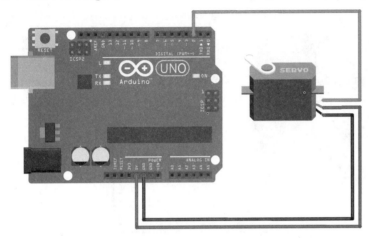

Figure C-14 Circuit diagram for servo motor

The code to control a servo with PWM is as below:

```
#include<Servo.h>

Servo servo_1;   // Create a servo object

int servo_pin = 3;   // PWM pin for servo control
int pos = 0;         // servo starting position

void setup() {
   servo_1.attach(servo_pin); // attach the servo_pin to the
                              //servo
}

void loop(){
   servo_1.write(180); // goes to 180 degrees
   //delay 1 second to allow the servo to reach the desired position
   delay(1000);
   servo_1.write(0);  // goes to 0 degrees
   //delay 1 second to allow the servo to reach the desired position
```

```
    delay(1000);
}
```

You can also use a potentiometer to make a position control. The Arduino will read the voltage on the middle pin of the potentiometer and adjust the position of the servo motor shaft (see Figure C-15).

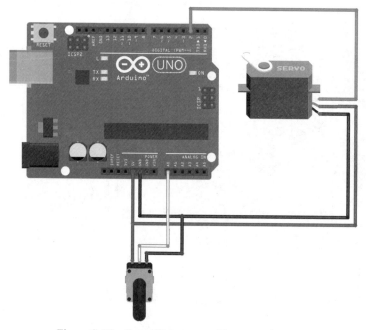

Figure C-15   Controlling servo with a potentiometer

The code below shows how to control a servo with a potentiometer.

```
#include<Servo.h>  //Servo library

Servo servo_test;  //initialize a servo object for the connected servo

int angle = 0;
int potentio = A0; // initialize the A0analog pin for potentiometer

void setup()
{
    servo_test.attach(3); //attach the signal pin of servo to pin 3
}
```

```
void loop()
{
    // reading the potentiometer value between 0 and 1023
    angle = analogRead(potentio);
    // scaling the potentiometer value to angle value between
    // 0 and 180
    angle = map(angle, 0, 1023, 0, 179);
    //command to rotate the servo to the specified angle
    servo_test.write(angle);
    delay(5);
}
```

```
#include<Servo.h>

Servo servo_1; //servo controller(multiple servos can be controlled)

int servo_pin = 3;   // PWM pin for servo control
int pos = 0;         // servo starting position

void setup() {
  servo_1.attach(servo_pin); // attach the servo_pin to the  servo
}

void loop() {
    for(pos=0; pos<=180; pos++) { // goes from 0 degrees to 180
                                  // degrees
        // in steps of 1 degree
        servo_1.write(pos); //tell servo to go to 'pos'
        delay(20); //delay to allow the servo to reach the desired
                   //position
        if(pos==90)
            delay(2000); //wait 2 seconds once positioned at 90
                         //degrees
    }

    delay(2000); // wait 2 seconds after reaching 180 degrees

    for(pos=180; pos>=0; pos==) { // goes from 180 degrees to
                                  //0 degrees
      servo_1.write(pos);//tell servo to go to 'pos'
      delay(20);
```

```
            if(pos==90)
                delay(2000); //wait 2 seconds once positioned at
                             //90 degrees
    }
    delay(2000); //wait 2 seconds after arriving back at 0 degrees
}
```

You are required to add more component to the circuit and modify the code.

Appendix: Vocabulary for C/C++

A

abstract class 抽象类
access 存取，访问
access function 访问函数
action 动作
activate 激活
active 活动的
actual parameter 实参
address 地址
algorithm 算法
alias 别名
align 排列，对齐
angle bracket 尖括号
annotation 注解，评注
application 应用程序
application framework 应用程序框架
appearance 外观
append 附加
argument 参数(传给函数式的值)，参见 parameter
array 数组
arithmetic 算术
arrow operator 箭头操作符
assembly 装配件，配件
assembly language 汇编语言
assign 赋值（动）
assignment 赋值，分配
associated 相关的，相关联的
attribute 特性，属性

B

backup 备份
base class 基类
binary 二进制
binary operator 二元操作符
bit 位
block 块，区块，语句块
bounds checking 边界检查

C

call 调用
character 字符
class 类
class declaration 类声明
class definition 类定义
class library 类库
class template 类模板
comment 注释
compile time 编译期，编译时
compiler 编译器
console 控制台
constant 常量
construct 构件，成分，概念，构造（for language）
constructor 构造函数，构造器
custom 定制，自定义

D

debug 调试
declaration 声明
default 缺省，默认值

definition 定义
dereference 解引用
derived class 派生类
design pattern 设计模式
destroy 销毁
destructor 析构函数
directory 目录
document 文档
dot operator (圆)点操作符
dump file 转储文件

E

encapsulation 封装
enum (enumeration) 枚举
equal 相等
equality 相等性
equality operator 等号操作符
escape code 转义码
escape character 转义字符
evaluate 评估
exception 异常
exit 退出
export 导出
expression 表达式

F

facility 设施，设备
field 字段(Java)
file 文件
firmware 固件
flag 标记
font 字体
framework 框架
full specialization 完全特化
function 函数
function call operator (即 operator()) 函数调用操作符
function object 函数对象
functionality 功能
function template 函数模板

G

getter (相对于 setter)取值函数
global 全局的
global object 全局对象
GUI 图形用户界面

H

header file 头文件
heap 堆
hierarchy 层次结构，继承体系
high level 高阶，高层

I

icon 图标
identifier 标识符
index 索引 (for database)
implement 实现
implementation 实现，实现品
implicit 隐式
import 导入
inheritance 继承，继承机制
inline 内联
initialization 初始化（名）
initialization list 初始化列表，初始值列表
initialize 初始化（动）
inner join 内联接 (for database)
instance 实例
integrate 集成，整合
interacts 交互
interface 接口
interpreter 解释器
iterate 迭代

iterative 反复的，迭代的

L

level 阶，层例
library 库
lifetime 生命期，寿命
link 连接，链接
list 列表，表，链表
load 装载，加载
local 局部的
login 登录
loop 循环
lvalue 左值

M

machine code 机器码，机器代码
macro 宏
member 成员
member function 成员函数
memory 内存
memory leak 内存泄漏
menu 菜单
method 方法
modifier 修饰字，修饰符
module 模块
multi-tasking 多任务

N

namespace 名字空间，命名空间
native code 本地码，本机码
nested class 嵌套类

O

object 对象
object file 目标文件
object oriented 面向对象的
operand 操作数
operation 操作

operator 操作符，运算符
option 选项
overload 重载
overloaded function 重载函数
overloaded operator 被重载的操作符
override 覆写，重载，重新定义

P

parameter 参数，形式参数，形参
parameter list 参数列表
parameterize 参数化
parent class 父类
parentheses 圆括弧，圆括号
parse 解析
parser 解析器
pass by address 传址（函数参数的传递方式）
pass by reference 传引用
pass by value 按值传递
pattern 模式
platform 平台
pointer 指针
polymorphism 多态
pop up 弹出式
postfix 后缀
prefix 前缀
preprocessor 预处理器
primitive type 原始类型
procedure 过程
procedural 过程式的，过程化的
process 进程
profile 轮廓，概要
program 程序
programmer 程序员
programming 编程，程序设计
project 项目，工程

property 属性

Q
qualifier 修饰符
queue 队列

R
random number 随机数
raw 未经处理的
readonly 只读
recursive 递归
refer 引用，参考
reference 引用，参考
register 寄存器
regular expression 正则表达式
relational database 关系数据库
return 返回
return type 返回类型
return value 返回值
robust 健壮的
robustness 健壮性
runtime 运行期，运行时
rvalue 右值（参考 value）

S
save 保存
scroll bar 滚动条
scope 作用域，生存空间
SDK (Software Development Kit)软件开发包
sealed class 密封类
search 查找
semantics 语义
serial 串行
serialization/serialize 序列化
specialization 特化

specification 规范，规格
source code 源码，源代码
standard library 标准库
standard template library 标准模板库
statement 语句，声明
stream 流
string 字符串
subobject 子对象
subroutine 子例程
subscript operator 下标操作符
subset 子集
subtype 子类型
symbol 记号
syntax 语法

T
table 表（for database）
template 模板
template parameter 模板参数
temporary object 临时对象
text 文本
text file 文本文件
thread 线程
token 符号，标记，令牌（看场合）
type 类型

U
UML (unified modeling language)统一建模语言
unary operator 一元操作符
user interface 用户界面

V
value types 值类型
virtual function 虚函数

References

[1] Prata S. C++ Primer Plus[M]. 6th Ed. 北京：人民邮电出版社，2019.
[2] Savitch W. Problem Solving with C++: The Object of Programming [M]. 10th Ed. 北京：清华大学出版社，2018.
[3] 顾春华. 程序设计方法与技术：C 语言[M]. 北京：高等教育出版社，2017.
[4] 史济民. 软件工程：原理、方法与应用[M]. 3 版. 北京：高等教育出版社，2010.
[5] Savitch W. C++入门经典[M]. 10 版. 北京：清华大学出版社，2018.